C0-ASY-865

Sputnik to Space Shuttle

by IAIN NICOLSON

DODD, MEAD & COMPANY *New York*

Published by Dodd, Mead & Company, Inc.
79 Madison Avenue, New York, N.Y. 10016
Manufactured in the United States of America
First Edition

Originally published in Great Britain
by Sidgwick & Jackson Limited, 1982

Library of Congress Cataloging in Publication Data

Nicolson, Iain.
Sputnik to space shuttle.

Bibliography: p.
Includes index.
1. Astronautics—Popular works. I. Title.
[TL793.N54 1985] 629.4′09 84-24716
ISBN 0-396-08231-9 (pbk.)

Acknowledgements

The author wishes to express particular thanks to the following for assistance in obtaining photographs: Mrs Sadie Alford (Novosti Press Agency), Mrs Gee (British Aerospace), Patrick Moore, Mat Irvine, and to NASA for its ready supply of information and illustrative material. Thanks are also due to John Hardwick for drawing the line diagrams, and to Bill Procter of Sidgwick and Jackson Limited for his assistance and encouragement.

Contents

	Acknowledgements	5
1	Sputnik—Herald of the Space Age	9
2	The Earth in Space	18
3	Ways and Means	28
4	Spaceprobes to the Moon	51
5	Robot Explorers of the Solar System	69
6	New Windows on the Universe	94
7	Satellites at Work	106
8	Man in Space	122
9	'One Small Step . . .'	137
10	Space Stations and Space Meetings	154
11	Into the Shuttle Era	168
12	The Way Ahead	181
13	Recent Developments – April 1982 to September 1984	193
	Appendix: Highlights of Space Activity	211
	A Guide to Further Reading	218
	Index	221

1

Sputnik–Herald of the Space Age

The Space Age began on 4 October, 1957. On that date the Soviet Union successfully placed Sputnik 1, the world's first artificial satellite, into orbit round the Earth. The 84-kilogram sphere, 58 centimetres in diameter, travelled round the Earth in a period of 96 minutes, its altitude ranging between 229 and 947 kilometres, and all the while its battery-powered radio transmitter emitted the characteristic 'bleep, bleep' signal so vividly imprinted on the minds of all who recall the day the Space Age dawned.

Throughout the world, the reaction to this event was one of astonishment. Although it had been known for more than two years that the United States was planning to launch small man-made satellites into space as part of their contribution to the International Geophysical Year (IGY), a 64-nation cooperative investigation into the Earth and upper atmosphere, the fact that the Soviet Union should have achieved this goal in advance of the Americans, and without a prior fanfare of publicity, came as a complete and utter surprise to the western world. The impact of Sputnik 1 on the minds and imagination of men was greater than that of any scientific event before or since. Across the globe, newspapers carried banner headlines announcing the event. Typical of the press reaction was the *Daily Express* headline on the following morning:

SPACE AGE IS HERE

The first 'Flying Saucer' travels at 17,000 m.p.h.

Almost immediately, speculation began as to the future course of space exploration. The *Sunday Express* of 6 October maintained that 'Leading authorities claim that they (the Russians) can send an expedition to the Moon in the next three to eight years and get space ships to Mars and Venus within five to ten years.' In the event, these claims proved to be premature and over-optimistic, but with unmanned space-probes, these goals were at least partially attained well within schedule.

The *Sunday Express* captured the essence of western reaction to the Soviet achievement in its editorial:

> 'In one dramatic move, the Russians have proved to the world the brilliance of their scientists and the genius of their technicians . . . In America understandably there is concern among ordinary people at the way in which America should apparently have been left behind.'

In the western world, and in the United States in particular, admiration for the great technical achievement was tinged with more than a hint of anxiety about the clear implications spelled out by the Russian ability to launch an 84-kilogram satellite when the Americans had not yet achieved their goal of launching one of less than 10 kilograms. Quite clearly the Russians had used a full-scale intercontinental ballistic missile (ICBM) to launch their satellite, and the capabilities of this launch vehicle were further emphasized on 3 November when Sputnik 2 was launched; this second satellite weighed no less than 500 kilograms. To place so heavy a satellite in orbit required a launch vehicle at least twenty-five times more powerful than the rocket which the Americans were developing to launch their first satellite, and several times more powerful than the ICBMs being developed in the United States at that time, and which were still suffering major teething problems.

The key to the Soviet success was their powerful launch

vehicle (see Chapter 3), brainchild of rocket engineer Sergei Korolev, backed by the flamboyant Soviet leader, Nikita Khrushchev. That particular rocket, and its derivatives, has remained the backbone of the Soviet space programme to this day.

To a world polarized by the east-west ideological conflict, the message was clear. The Soviet Union, up to that time considered to be technologically backward compared to the United States, had taken the lead in the newest and most spectacular frontier of exploration and scientific endeavour and – on a more sinister level – clearly possessed the means to deliver nuclear warheads directly from Soviet soil to the United States. The almost axiomatic belief widely held in the United States, of American military and technological superiority over all other nations, was shattered at a stroke. The world at large was deeply impressed. Soviet success in space was held up as a clear indication of the superiority of the socialist system over the capitalist system, and this theme was vigorously pursued in the early days of the Space Age as the Soviet Union pulled off a string of spectacular space 'firsts'.

Despite official efforts to play down the significance of Sputnik – President Eisenhower remarked that 'The Russians have only put one small ball in the air' – the United States administration was clearly worried both by the world-wide reaction to the launching of Sputnik and by the strategic implications of its launch vehicle. The missile programme and the space programme were re-assessed, and accelerated. The launching of Sputniks 1 and 2 followed on 6 December, 1957 by the spectacular failure of the first American attempt to launch a satellite brought American self-esteem to a very low ebb, but acted too as a powerful catalyst to the development of the American space programme, a programme which would lead – within twelve years – to the first Moon landings. America took up the challenge. Despite occasional official denials of its existence, the 'Space Race' was most definitely on.

Although the launching of Sputnik came as such a great surprise, the Soviet intent to launch satellites had been made plain beforehand on a number of occasions. During the late 1940s and early 1950s there had been considerable general discussion about the feasibility of launching satellites and of their potential scientific and military value. On 29 July 1955, President Eisenhower formally announced that the United States would launch a number of small unmanned Earth satellites during the International Geophysical Year which was due to commence on 1 July, 1957. Just a few days later, at the Sixth Congress of the International Astronautical Federation, the Russian Professor of Physics, Leonid Sedov stated that the Soviet Union would also launch satellites during the IGY, and that they would be larger than those which the Americans were proposing. This statement was largely ignored in the West at the time, as was the earlier announcement in the Soviet press on 4 April 1955, that an Interdepartmental Commission on Interplanetary Communications had been set up to develop satellites.

The official announcement of Soviet intent to launch satellites during IGY was made in September 1956 at a conference held in Barcelona in preparation for the IGY. In June 1957 the Soviet press announced that the Soviet Union would launch its first satellite within a few months, and even gave the radio frequencies – 20 and 40 MHz – at which it would be transmitting. Even at this stage, little attention was paid or credence attached to such statements outside of the Soviet Union.

Thus it was that when the first Sputnik was blasted into orbit on 4 October, to all but a few *cognoscenti,* the announcement of the event came as a bolt from the blue. The launching had taken place at 19.12 hours G.M.T. from a site now known officially as the Baikonur Cosmodrome, described in Soviet statements as being near the town of Baikonur, but in fact located some 370 kilometres south-west of Baikonur, near the town of Tyuratam, east of the Aral sea in the province of Kazakhstan. The satellite itself remained in orbit for 92 days. The small but finite drag exerted by atmospheric friction when the satellite was at its lowest point in its orbit – a point known as perigee – inexorably lowered the orbit until, on 4 January, 1958, Sputnik 1 re-entered the atmosphere and burned up. Although it ceased to transmit after three weeks, valuable information about the density of the upper atmosphere was gained from tracking the satellite and observing the way in which atmospheric drag modified its orbit. The satellite itself was not bright enough to be seen without optical aid, but the spent final stage of the rocket, which also went into orbit, could be seen quite readily as a point of light slowly moving against the background stars. Today there are so many satellites in orbit that their sighting in the sky is commonplace and unremarkable, but in the first few years of the Space Age, they aroused the widest interest, and national newspapers used to carry regular bulletins on which satellites would be visible at what times of the night. For a time at least, people the world over were moved to cast their eyes skywards and search for the moving points of light which symbolized the fact that the Space Age had arrived.

The much more massive Sputnik 2 went into orbit on 3 November. The satellite consisted of two principal modules, a communications section, and a pressurized section containing the dog 'Laika' – the first living creature to go into space. Laika survived the acceleration forces involved in being blasted into space, and endured the weightless condition in orbit without suffering ill effects. However, she could not be returned to Earth as the Sputnik possessed neither the retro-rocket necessary to slow down its motion and bring it back down into the atmosphere, nor the heat shield necessary to survive the process of re-entry. About ten days after entry into orbit, the oxygen supply ran out, and Laika died in space. Worldwide reaction to the sacrifice of the dog in this mission was mixed, but the fact remains that Laika's spaceflight was the first crucial step towards manned spaceflight which was to follow.

Sputnik 2 survived in orbit for 162 days, re-entering and burning up on 14 April, 1958.

In the United States the first serious proposal to place a satellite in orbit was made in 1954 by a U.S. Army team headed by Wernher von Braun. Project Orbiter, as it was known, was intended to use the Army's Redstone intermediate range missile as the first stage, and solid-fuelled rockets as upper stages, to place in orbit a satellite weighing only a few kilograms. Given the go-ahead, von Braun was confident of being able to launch a satellite by January 1957.

President Eisenhower, however, was determined that the

satellite programme be seen clearly to be separate from the military missile programme. Accordingly Project Orbiter was rejected in favour of one of two proposals made by the U.S. Navy. Project Vanguard, officially sanctioned in 1955, was based on a new launch vehicle to be developed solely for the scientific satellite programme, and without military application. The first stage of this new rocket was to be a modified Viking rocket, a type of rocket which had been developed and used purely for upper atmosphere research. The term 'Vanguard' was applied both to the launch vehicle and to the satellites which were to be placed in orbit.

From its inception, Project Vanguard was widely publicized, and many articles appeared describing in detail what was confidently expected to be the world's first artificial satellite. Problems were encountered in developing the launch vehicle, and the programme slipped behind schedule. Nevertheless, after the launching of Sputnik 1, the programme was accelerated, and on 6 December, 1957 what was originally intended as a test launch of the Vanguard vehicle was, by pressure of events, upgraded to a full-scale attempt to launch America's first satellite – a package of instruments about the size of a grapefruit. Television crews and newsmen descended on Cape Canaveral in droves. The launching was to be shown live on national television; in the public mind, this was to be the event which retrieved American pride, so badly dented by the two Sputniks. In the event, the Vanguard rocket rose to a height of about one metre, exploded in a brilliant ball of flame and smoke, then collapsed ignominiously onto what was left of the launch pad. The failure assumed the proportions of a national disaster. The second Vanguard launch attempt on 5 February, 1958 was only marginally less disastrous – the rocket breaking up at an altitude of about 5 kilometres, but by that time American prestige had been salvaged from a completely different direction.

On 8 November, 1957, just five days after the launching of Sputnik 2, official approval was given to von Braun's team to use a four-stage modified version of their Jupiter-C launch vehicle – itself based on the Redstone, and already proven in a number of long-range test firings – to place a satellite into orbit. On 31 January, 1958, just twelve weeks after the project had been given the go-ahead, and only one day later than scheduled, the Army launcher – now designated Juno-1 – successfully fired Explorer 1, the first American satellite, into orbit. Explorer 1, weighing just 14 kilograms, was placed in an orbit which ranged in altitude between a minimum of 356 kilometres and a maximum of 2548 kilometres – much higher than the earlier Sputniks.

Explorer's instrumentation included a Geiger counter, designed to detect 'cosmic rays' (high energy atomic and nuclear particles which were known to be reaching the upper atmosphere from the depths of space and whose origin, to some extent, remains a mystery to this day). As expected the counts of particles registered on the detector increased with altitude, but as Explorer 1 passed above about 1000 kilometres the counts seemed to stop completely. The problem was resolved by Professor James van Allen who concluded that, in fact, beyond this level the numbers of particles were so high that the instrument became completely saturated.

Explorer 3, launched on 26 March into an orbit which reached a maximum altitude of 2800 kilometres, fully confirmed this conclusion, as did Sputnik 3, two months later.

What Explorer 1 had discovered was the innermost of two zones, known as the Van Allen zones, within which electrically charged particles are trapped by the Earth's magnetic field. The inner zone, containing mainly protons, has its greatest concentration at a height of some 3000 kilometres and the outer zone, mainly electrons, is centred on 15,000 to 20,000 kilometres altitude. The outer zone was discovered by the American space probe Pioneer 3, in December 1958. Within these zones charged particles are forced to oscillate to and fro in the north-south direction between the polar regions as a result of their interaction with the lines of force of the Earth's magnetic field.

Thus, the first major scientific discovery of the Space Age had fallen to the tiny American Explorer, launched by a stop-gap vehicle in response to the challenge posed by the first two Sputniks.

The existence of these hitherto unsuspected radiation belts was to be a source of anxiety for some time to come with regard to the possible harm which might befall future astronauts who travelled into these regions of space. In the event, the Van Allen zones have not impeded manned space flight, but in those early days, the discovery prompted genuine concern.

Somewhat belatedly, the Vanguard programme met with its first success when the tiny 1.5-kilogram payload of Vanguard 1 was fired into an elliptical orbit ranging in height from 650 to 3970 kilometres. Study of the motion of this tiny satellite revealed the fact that the Earth is not spherical, but slightly pear-shaped. Subsequent studies of the motion of satellites have been instrumental in further refining our knowledge of the shape and internal structure of the Earth. Nine days later, Explorer 3 was safely inserted into orbit, Explorer 2 having failed to reach orbit. The scientific dividends from the two Explorers and the first Vanguard was considerable.

Just as America was beginning to taste the fruits of success in space exploration, there came the launching of Sputnik 3 – by far the heaviest satellite yet to be launched. Propelled into orbit on 15 May, 1958, the satellite weighed 1327 kilograms, and of this some 970 kilograms consisted of scientific instrumentation for measuring solar radiation, cosmic rays, the radiation belts, and the properties of the upper atmosphere. The great weight of the satellite testified to the power of the launch vehicle, and hinted at the possibility that this vehicle could be used to place a man into orbit.

By the time the first anniversary of the launching of Sputnik 1 came round, the tally of successful satellite launchings amounted to three for the Soviet Union (the Three Sputniks) and four for the United States (three Explorers and one Vanguard). The Soviet score card recorded the first ever artificial satellite, the first living creature to be placed in space, and the heaviest satellite (indeed, even Sputnik 1 was heavier than the combined weight of all four American satellites). While the Americans justifiably could claim the first really significant scientific results, in the public eye, this could by no means outweigh the impact of the space 'firsts' achieved by the Sputniks.

History may count it a tragic reflection on our times that mankind's breakout into space should have occurred in the context of a contest between the two superpowers, each attempting to demonstrate its technological prowess to the rest of the world, and that the rocket vehicles which rendered this possible should have been developed as the means to hurl weapons of mass destruction across the globe. But the history of mankind has been one of conflict, and conflict – real or hypothetical – has provided a spur to technological innovation; the whole history of the rocket has been inextricably entwined with warfare. Indeed, the 'Space Race' which followed the launching of Sputnik 1 may well have played a vital role in alleviating the tensions of East-West conflict by providing a grand arena for the competitive display of technological achievement.

Be that as it may, in the long term, the real significance of Sputnik 1 is that it represents the first step in casting off the gravitational shackles which have tied our species to one small planet – Earth. We have set forth on a road which has already allowed us to begin to achieve the centuries' old dream of travelling through space and visiting other worlds. Since 4 October, 1957, mankind's space activity has blossomed forth in multifarious directions: men have been to the Moon; unmanned spacecraft have explored many of the planets; instruments aboard orbiting satellites have plumbed the depths of the universe in ways which would have been unattainable by Earth-bound astronomers; satellites have greatly enhanced our knowledge of the Earth itself; and satellites have been put to work in the service of mankind – for communications, meteorology, surveying, and a host of other applications.

The chapters which follow tell the story of the first quarter century of the Space Age, its triumphs, failures, discoveries, and achievements. It is a story which is only just beginning.

2

The Earth in Space

Man is an inquisitive creature with an insatiable desire to explore his environment, and Man's environment, in its widest sense, is the universe, a boundless arena for physical and intellectual exploration. Until the dawn of the Space Age, Man was restricted to the surface of the Earth and could do no more than view other worlds through his telescopes; now we have begun the physical exploration of our nearest neighbours in space and there seems to be no natural reason why we should not eventually extend our frontier of exploration far beyond the nearby planets.

For the present, where do we stand in relation to the universe?

The Earth-Moon System

We live on the planet Earth, a small body which travels round a star – the Sun – in an elliptical path, or orbit, at a mean distance of 149,600,000 kilometres. This distance is known as one *astronomical unit*. The Earth is a relatively dense body, made up mainly of metals, such as iron and nickel, and silicates – rocky materials. Deep in the interior, where the temperature probably exceeds 5000°C, is the liquid metallic core; this is overlaid by a mantle of dense rocks some 2,900 kilometres thick, and above this in turn lies the less dense crust, at most some 40 kilometres thick. The Earth is unique among the known planets in that about 70% of its surface is covered in water, oceans of this vital liquid as deep as ten kilometres in places. The mean radius of our planet is about 6400 kilometres; the continental land masses rise above the mean ocean level to a maximum height of about 8 kilometres in the highest mountain ranges.

Vital for the existence of life on Earth is the atmosphere, an envelope of gas composed primarily of nitrogen (78%) and oxygen (21%), which provides air for us to breathe and which protects living organisms from harmful radiation. Minor constituents play a crucial role. For example, ozone (a form of oxygen molecules made up of three atoms) comprises less than one millionth of the mass of the atmosphere, yet is very largely responsible for absorbing ultraviolet radiation from the Sun which, if it were to penetrate to ground level, would have a deadly effect on living tissue. The atmosphere also acts as a blanket, preventing heat from escaping easily into space, and maintaining a higher, more uniform, global temperature than would be the case if the atmosphere were absent. The oceans, the land masses, and the lower reaches of the atmosphere are teeming with life of a bewildering variety of form.

Our nearest celestial neighbour is the Moon, the Earth's natural satellite. Travelling round the Earth at a mean distance of 384,000 kilometres, it is a barren and hostile world; about one quarter of the size of the Earth, it is devoid of oceans and atmosphere, and its surface is pitted with craters. Nevertheless, it is important to us, providing light at night, being largely instrumental in raising the ocean tides, and it represents the only other world on which men have set foot, so far.

The Sun

The Sun is of crucial importance to us. It is the source of light, heat, and most forms of energy here on Earth; without the Sun we simply should not exist. So far as we can tell, the Sun is a very ordinary, middle-of-the-road star, quite unexceptional except for being very much nearer to us than any other star. Like other stars it is a self-luminous globe of hot gas, producing energy by means of nuclear reactions – similar to those which take place in the hydrogen bomb – in its central core. The Sun is composed mainly of the two lightest chemical elements, hydrogen and helium; the other elements – more familiar to us here on Earth – comprise only about 2 per cent of the Sun's mass. With a radius of just under

700,000 kilometres it is over a hundred times larger than our planet, and its globe could contain well over a million bodies the size of Earth. Its mass is about 330,000 times greater than that of the Earth, but its mean density is only about a quarter of the terrestrial value.

The visible surface is known as the photosphere (literally 'sphere of light') and has a temperature of just under 6000 K*. Deep in the interior, the temperature is believed to be about 14,000,000 K, and at this very high temperature the process of *fusion* takes place, whereby the lightest element, hydrogen, is converted into the next element, helium, with the release of copious quantities of energy. In each nuclear reaction a small quantity of matter is turned into energy; to keep the Sun shining at its present level, over four million tonnes of matter are converted into energy every second. This rate of mass loss is of little consequence to the Sun. It has been shining for about five billion years, and we believe that it has sufficient reserves of fuel to keep it going at its present rate for a further five or six billion years. Before the Sun finally runs out of fuel and shrinks under its own weight to form a dense, compact body known as a white dwarf, it will swell temporarily and increase in brightness several hundred-fold, to become a type of star known as a red giant. When this happens the Sun will expand sufficiently to engulf the planet Mercury while here on Earth the oceans will evaporate, the atmosphere will be driven off into space, the surface will become searingly hot, and life on Earth will become impossible. If any descendants of the human race exist on this planet when that time comes, they will have to remove themselves to another world where conditions are more equable – or perish. In the very long term, space travel will become essential to the survival of the species!

The Solar System

The Sun is the central, dominant member of the Solar

* In the International System of Units which is gradually becoming adopted in scientific circles, the unit of temperature is the *kelvin*, denoted by 'K'. The Kelvin, or Absolute, temperature scale begins at absolute zero, a temperature of −273 degrees on the centigrade scale. To convert from centigrade to kelvin, *add* 273; to convert from kelvin to centigrade *subtract* 273. With the high temperature of stars, the difference between kelvin and centigrade temperatures is of little consequence.

System, a system of bodies which comprises the Sun, nine known planets, the planetary satellites, a host of minor bodies – asteroids, meteoroids, and comets – and a certain amount of gas and dust.

The planets, in order of distance from the Sun, are: Mercury, Venus, Earth, Mars, Jupiter, Saturn, Uranus, Neptune, and Pluto. The first four are similar to the Earth in the sense that they are relatively small dense bodies with solid surfaces; they are known as the terrestrial planets. Their distances from the Sun (see Table 1) range from 0.4 astronomical units for Mercury to just over 1.5 astronomical units for Mars. Beyond Mars there is a broad gap, the next planet, Jupiter, lying at just over 5 astronomical units from the Sun.

Jupiter is the first of the four giant planets, and is the largest of all the planets in the system. Eleven times the size of the Earth and 318 times as massive, it is truly a giant world. It is utterly different from the Earth. Composed mainly of hydrogen and helium (its composition is much more like that of the Sun than that of the Earth), it has no solid surface; below its deep gaseous atmosphere, the interior is mainly liquid hydrogen. There may be a dense, rocky-metallic core right at the centre, but we have no firm evidence as to its size.

Saturn, nearly twice as far from the Sun as Jupiter, is slightly smaller and considerably less massive. Uranus and Neptune, both about four times larger than Earth are smaller than Jupiter and Saturn, but a little denser. Saturn is surrounded by a beautiful and amazingly complex system of rings (see Chapter 5), and both Jupiter and Uranus are known to have much more modest rings; there is every reason to suppose that Neptune, too, will turn out to have a ring system.

Pluto is a curious world. Discovered as recently as 1930, it is tiny, no larger than Mercury and probably not much larger than our own Moon. It seems to have a very low density and it may be an icy world – a cosmic snowball. Pluto's orbit is markedly elliptical, at its closest approach to the Sun (a position known as 'perihelion') its distance is 29.6 astronomical units, but at its greatest distance ('aphelion') it is 49.2 astronomical units distant from the Sun. When Pluto is near to perihelion, it is closer in than Neptune; this is the situation at present, and will remain so until the end of this century, as Pluto slowly traces out its distant orbit – taking 248 years to complete each circuit of the Sun.

Mercury and Venus have no natural satellites (moons), the Earth has 1, Mars has 2 tiny moonlets, and the giant planets have a multitude – Jupiter having at least 16 and Saturn more than 20. So far as is known at present, Uranus has 5 and Neptune 2. Even tiny Pluto has a satellite.

Minor members of the Solar System

Between the orbits of Mars and Jupiter lie thousands of tiny bodies – the asteroids, or minor planets – each pursuing its separate orbit round the Sun. The first of them was discovered in 1801 and since that time over two thousand have been studied in sufficient detail to have their orbits determined. In all, there must be at least a hundred thousand asteroids, ranging in size from just under 1000 kilometres in diameter, down to irregular rocky bodies less than 1 kilometre across. Although the great majority lie between Mars and Jupiter, some have been found beyond Jupiter and Saturn, while others move in orbits which pass inside that of the Earth (the so-called 'Apollo' asteroids) and even within the orbit of Mercury Whether the asteroids represent the fragments of a former planet, or simply debris from the formation of the Solar System is not known for sure, although the latter view seems much more likely.

In the not-too-distant future, the asteroids may prove to be of great significance to us; future generations may mine them for their mineral wealth.

The most ethereal, and occasionally most spectacular members of the Solar System are the comets. A typical bright comet consists of a compact nucleus surrounded by a cloud of gas and dust known as the coma from which extends a tail which stretches for millions of kilometres. (The tails of some of the greatest comets have extended for distances greater than the distance from the Sun to the Earth.) Generally, there are two tails, one composed of dust and the other of ionized gas (gas comprising atoms which have been stripped of one or more of their electrons). Both types of tail are driven from the head of the comet by solar radiation, the gas tail being expelled by the action of the solar wind – the stream of atomic particles which blow away from the Sun – and the dust tail by the pressure exerted by light itself. Most comets follow long, elongated elliptical orbits which bring them periodically from the outermost depths of the Solar System to a close encounter with the Sun. The tail develops as the comet approaches the Sun and declines as it recedes; the tail, or tails, always point more or less directly away from the Sun whether the comet is approaching or receding.

They are fascinating objects, and plans are afoot to send probes to the most famous of them – Halley's comet – on its next return in 1986.

The Stars

Far beyond the fringe of the Solar System lie the stars. Interstellar distances are so vast that it is almost meaningless to think of them in terms of conventional units of measurement; when we start talking about millions of millions of kilometres we can no longer even attempt to visualize such distances. However, a useful indication of relative distances can be obtained by thinking of how long it would take for a ray of light, or a radio signal to cross these distances. Light and radio waves are both forms of electromagnetic radiation (see Chapter 6) and travel through space at a speed – known as the speed of light – of 300,000 kilometres per second. Light is the fastest-moving entity in the universe. Light travels a distance of 9.5 million million kilometres in a year, and this distance is known as a *light-year*.

In these terms, the nearest star – a dim red star known as Proxima Centauri (too faint to be seen without telescopic aid)– lies at a distance of 4.2 light-years; in other words, a ray of light requires 4.2 years to cover the distance from Proxima Centauri to the Earth. By way of comparison, light crosses the distance from the Moon to the Earth in 1.3 *seconds*, reaches us from the Sun in 8.3 *minutes*, and requires just over 5 *hours* to reach Earth from the outermost planet Pluto. Clearly the gulf which separates us from the stars is vast

TABLE 1

Major Bodies of the Solar System

Body	*Mean distance from the Sun*		*Orbital Period*	*Mass*	*Equatorial radius*	*Mean density*	*Axial rotation period*
	astronomical units	millions of km		(Earth = 1)	(km)	(water = 1)	
The Sun				330,000	696,000	1.41	25d
Mercury	0.39	57.9	88.0d	0.055	2,440	5.5	58.6d
Venus	0.72	108.2	224.7d	0.82	6,050	5.2	243d
Earth	1.00	149.6	365.3d	1.00	6,378	5.5	23h 56m
Mars	1.52	227.9	687.0d	0.11	3,397	3.9	24h 37m
Jupiter	5.20	778.3	11.86y	317.9	71,400	1.3	9h 55m
Saturn	9.54	1,427.0	29.46y	95.2	60,000	0.7	10h 14m
Uranus	19.18	2,869.6	84.01y	14.6	25,600	1.7	23h (?)
Neptune	30.06	4,496.7	164.8y	17.2	24,800	1.8	22h (?)
Pluto	39.44	5,900	247.7y	0.002 (?)	1,500 (?)	?	6.4d

N.B. 1 astronomical unit = 149,600,000 kilometres
The mass of the Earth is 5.97×10^{24} kilograms (10^{24} is a shorthand way of writing 1 followed by 24 zeros)
In the above, d = day, y = year, h = hour and m = minute.
Several values for Pluto are very uncertain and are indicated by '?'

compared with interplanetary distances. In fact, Proxima Centauri is about a quarter of a million times further away than the Sun.

Stars differ widely from each other in their properties. Although many stars are similar to our Sun, there are some which are up to several hundred thousand times more luminous, and others – true celestial glow-worms – which emit only one hundred thousandth of the Solar luminosity. There are stars which are hotter, and stars which are cooler; although there are exceptions, beyond the normal range, most stars have surface temperatures of between 35,000 K and 2,500 K.

There are stars known as red giants, and supergiants, which are hundreds, or even thousands of times larger than the Sun. For example, the bright red star Betelgeuse, in the constellation of Orion, is so large that if it were to be placed where the Sun is, then all the planets out to and including Mars would be inside its globe. At the other end of the scale, there are dying stars, known as white dwarfs, which are comparable in size with the Earth, and fantastically collapsed objects, called neutron stars, which are less than 10 kilometres in radius. Whereas red supergiants are less dense than air, white dwarfs are so dense that a teaspoonful of their material would weigh several tonnes, and neutron stars are so compressed that a teaspoon laden with neutron star material would weigh about a billion tonnes!

Even the incredible density of a neutron star may not represent the greatest extent to which matter can be compressed. It is widely believed that there exist black holes – regions of space into which matter has fallen, and within which the force of gravity is so powerful, that nothing – not even light itself – can escape. One way in which black holes are thought to form is by the collapse of massive stars when, at the end of their lives, they run out of fuel and are no longer able to support their own weight. Such stars are thought to continue to collapse until all their matter is crushed to infinite density, but before they reach this stage the force of gravity at the surface of the collapsing star reaches a level which prevents light escaping, and the star disappears from view. A black hole, then, consists of a collapsed mass surrounded by a region of space out of which nothing can escape. The evidence for their existence is now quite good.

Stars are born in clouds of gas, some of which can be seen directly as luminous clouds, or nebulæ, if they have embedded within them very hot, highly luminous stars which cause the surrounding gas to emit light. There are many clouds which cannot be seen directly in this way, but they can be detected by means of radio telescopes; for example, clouds of hydrogen in space emit radio waves at a wavelength of 21 centimetres (i.e. a frequency of 1420 MHz). If a region in a cloud is sufficiently dense, it will collapse under its own gravitational attraction, becoming hotter and denser until nuclear reactions commence, and it becomes a stable star like the Sun. Eventually stars run out of fuel and die. There are three possible fates which can befall a star: if it is comparable with, or less massive than, the Sun, almost certainly it will become a white dwarf which, over billions of years will cool down and fade away; if it is more massive, it may become a neutron star. it seems likely that such stars suffer a violent explosive outburst – known as a supernova – which hurls much of the star's material into space while at the same time compressing its core into a neutron star. Finally, the most massive stars may collapse directly to form black holes.

The Galaxy

The Sun and the other stars visible in the night sky are part of a system of a hundred thousand million stars, known as the Galaxy. The Galaxy is a disk-shaped system, with a concentration of stars in a central nucleus, this nucleus being surrounded by a thin disk of stars and interstellar material. The disk of the Galaxy is about 100,000 light years in diameter, but the outermost fringes of the system may extend much further. The Sun, together with the Solar System, lies about 30,000 light years from the centre of the Galaxy – an inconspicuous star in an unexceptional part of the system.

The Universe

Beyond the confines of the Galaxy we can see billions of other galaxies. The nearest large galaxy, known as M.31 and lying in the constellation of Andromeda, is at a distance of some 2,200,000 light years; if you know exactly where to look, it is just perceptible to the naked eye, and is certainly the most distant object which can be seen without the aid of a telescope. Other galaxies can be detected out to ranges of billions of light years and there are objects known as quasars which are so luminous that they can be detected much farther off. The most distant quasar lies at a distance in excess of 15 billion light years; its light has been travelling towards us for a period of time far in excess of the age of the Sun and Earth.

This, then, is the scale of the universe at the limit of present-day observations. So far, mankind has taken only the first tiny steps in its exploration. For the moment, and for some time to come, we must confine our physical exploration to the confines of the Solar System; but if mankind survives the next few critical decades, there can be little doubt that it will turn its attention seriously to the question of exploring beyond the confines of our own planetary system into interstellar space. Just as the human race is now beginning to climb out of its cradle, the Earth, so eventually it will surely move beyond the nursery of the Solar System.

For the present we must be more modest in our ambitions. In the years which have elapsed since the launching of Sputnik 1, a vast amount has been accomplished. In the next chapter we shall see how this odyssey of discovery was made possible.

3

Ways and Means

To travel to the Moon, the planets, or beyond we must overcome the effects of the Earth's gravitational attraction. According to the Law of Universal Gravitation set out by Isaac Newton in 1687, each body in the universe attracts every other body, the force between two bodies depending upon their masses, and diminishing with the square of distance (i.e., if the distance is doubled, the force of attraction is reduced to one quarter of its previous value). The gravitational influence of the Earth never disappears completely, no matter how far away one travels, but for all practical purposes, by the time a spacecraft is, say, a million kilometres distant, the Earth's attraction is almost negligible.

Near the Earth's surface an object allowed to fall freely is accelerated towards the Earth's centre by the Earth's gravitational attraction at a rate of 9.8 metres per second per second (i.e. each second, the speed of the falling object increases by a further 9.8 m/s so that after the first second its speed is 9.8 m/s, after the second second it has increased to 19.6 m/s, and so on), this rate of acceleration being known as 1 g. Standing on the Earth's surface we are resisting this accelerating force and therefore feel the sensation of weight. Your weight is the magnitude of the force exerted by the Earth on your body; the more massive the person, the greater the weight. Weight is experienced only if we are resisting gravity; but if we allow outselves to fall freely, we experience no sensation of weight.

For example, if you were inside an elevator when the supporting cable snapped, both you and the elevator would fall freely down the lift-shaft, both accelerating at the same rate. As a result, you would float freely inside the elevator with no sensation of weight. No experiment carried out inside the elevator could prove that you and the elevator were falling freely under gravity; you could equally well be inside a box out in the depths of space, far from the gravitational attraction of a massive body. Of course, when the elevator struck the ground, the sensation of weightlessness would be removed – permanently.

The effects of gravity can be removed by falling freely. Likewise an effect identical to that of gravity can be produced by acceleration. Imagine an astronaut in a sealed box inside a spacecraft in the depths of space. If the rocket motor is firing to accelerate the spacecraft, he will feel himself pressed against the floor of his box, just as if he were standing on the surface of the Earth and subject to the Earth's gravitational attraction. No experiment which he can carry out inside the box will tell him whether he is inside a box resting on the Earth's surface, or inside a spacecraft with the motor firing. If the motor is accelerating the spacecraft at a rate of 9.8 metres per second per second (1 g) the sensation of weight experienced will be just the same as the astronaut's normal weight on Earth. If the acceleration is greater, the 'weight' is greater, too.

The greater the height from which an object is dropped, the greater the speed with which it strikes the ground. Conversely, the faster an object is thrown, the higher it will reach before falling back down to Earth; to reach a given height, an object must be thrown upwards with a speed exactly equal to the speed with which an object dropped from that height would strike the Earth. To be able to throw a spacecraft to an infinite distance, it must be given the same speed as it would attain if it fell to the Earth from an infinite distance. This speed can be calculated, and is known as the *escape velocity*. At the Earth's surface, escape velocity has a value of 11.2 kilometres per second (i.e. about 40,000 kilometres per hour). If a spacecraft is to escape from the Earth and continue to move away without ever falling back, this is the *minimum* speed at which it must be projected. In practice, a spacecraft fired at this velocity will be reduced to crawling pace – relatively speaking – by the time it has travelled a million kilometres from Earth; to travel out to the planets, it is necessary to fire the spacecraft somewhat faster than the minimum velocity.

A spacecraft does not need to travel at such a high velocity in order to remain in orbit round the Earth. A satellite following a circular orbit close to the Earth's surface has to move at circular velocity, 7.8 km/sec; the value of circular velocity diminishes with distance; thus a satellite at a distance of 42,000 kilometres from the centre of the Earth travels at only 2.9 km/sec (and travels round the Earth once every 24 hours), while the Moon, at a distance of 384,000 kilometres moves at a more leisurely pace of 1 km/sec.

Why does a satellite remain in orbit, and what prevents the Moon from falling on our heads? According to Newton's first law of motion a body continues to move at constant speed in a straight line unless acted on by a force. Consider what would happen if a body were thrown in a direction parallel to the ground from the top of a high tower (Fig. 1). If the Earth's gravitational attraction were 'switched off', the body would continue to move in that direction at a constant speed for ever more. The effect of the Earth's attraction is to accelerate the body in the direction of the centre of the Earth, and as a result, it will fall to the ground some distance from the foot of the tower; the faster it is thrown, the further it will travel before striking the ground. Since the Earth is round, there must be some velocity at which a projectile will travel right round the globe, returning to its starting point. The combination of the tangential ('sideways') motion and the radial ('falling') motion ensures that the body moves round the Earth at a constant altitude; i.e. it follows a circular orbit.

An increase in velocity beyond circular velocity causes the satellite to follow an elliptical path, the point of closest approach to Earth being called 'perigee' and the point at which its distance is greatest, 'apogee'. A further increase in velocity produces a more elongated ellipse, and an increase to the escape velocity causes the satellite to follow an open curve which never returns to the starting point; such a curve is called a parabola, and escape velocity is also known as 'parabolic velocity'. An increase beyond escape velocity places the spacecraft on a still more open curve – a hyperbola –

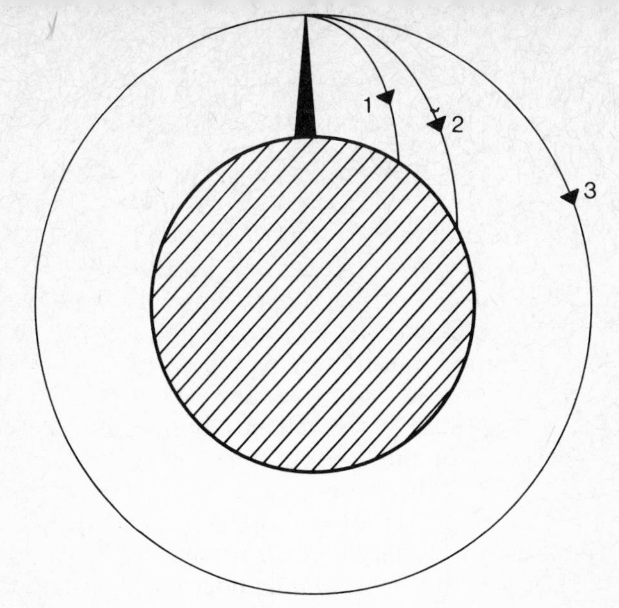

Figure 1.

(a) If a projectile is thrown horizontally from a high tower at a low speed, it will fall to the Earth's surface close to the base of the tower (1); if it is thrown at a higher speed it will travel a greater distance before striking the ground (2). If it is thrown at 7.8 km/sec (circular velocity), it will travel completely round the Earth in a circular path (3).

(b) If the speed of a satellite or spacecraft is increased beyond circular velocity it enters an elliptical orbit (4), the point of closest approach being *perigee* (P) and the point at which its distance is greatest being *apogee* (A). If the velocity is increased to 11.2 km/sec (escape velocity) the spacecraft will move away along a parabolic path (5), while if the speed is increased beyond this value it will follow a hyperbolic path (6).

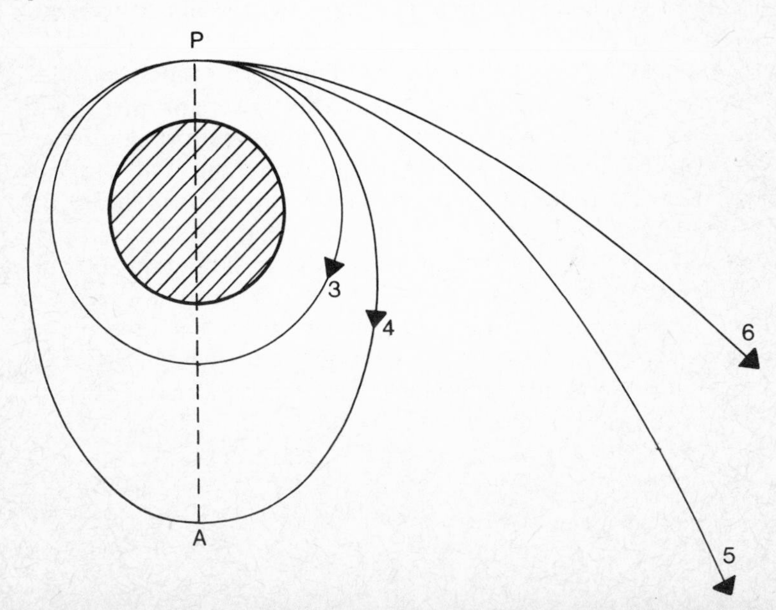

and the speed of the craft never drops to zero, settling instead to a constant value.

When a spacecraft is to be sent to the Moon or the planets, it is often placed into a circular or elliptical 'parking orbit' round the Earth before the rocket motor is fired again to accelerate the spacecraft to beyond escape velocity; it then arcs away from the Earth along a hyperbolic path.

Strictly speaking, the Moon can be reached by a spacecraft launched at just a little less than the escape velocity. There is a point – the neutral point – about nine tenths of the distance from the Earth to the Moon, at which the attractions of Earth and Moon are equal and opposite; if a spacecraft is given just enough velocity to take it past that point, it will leave the Earth's influence and fall under the attraction of the Moon. Provided it has been correctly aimed, the spacecraft will then fall directly towards the Moon, accelerating as it falls, to strike the lunar surface with a speed approximately equal to the Moon's escape velocity (2.4 km/sec; i.e. 8600 km per hour).

Although the basic theoretical knowledge necessary to work out the speeds and trajectories required for spaceflight has existed since the time of Newton, 270 years were to pass between the publication of his theory of gravitation and the placing of Sputnik 1 into orbit round the Earth. It took that length of time for technology to develop the means to travel at the seemingly awesome speeds necessary to achieve space travel. Before the development of modern launch vehicles, many ingenious schemes for 'space travel' were proposed.

The theme began to crop up in literature a surprisingly long time ago. The earliest known fictional account of a voyage to the Moon is the oft-quoted *True History*, written by the Greek satirist, Lucian of Samosata, in the second century A.D. The heroes of this delightful tale were sailing through the Pillars of Hercules (the Straits of Gibraltar) when their ship was picked up by a waterspout, hurled into the air and blown along for seven days and seven nights after which they were deposited, to their considerable surprise, on to the surface of the Moon. This turned out to be inhabited, and proved to be the scene of major battles between warring factions!

The seventeenth century, a period of great scientific and

cultural activity, saw the publication of a good number of fictional works with themes related to space travel. The great astronomer Johannes Kepler, who showed that the planets move round the Sun in elliptical orbits (rather than in circles, as Copernicus had believed) wrote a story called *Somnium* (The Dream) which concerned a rather mystical journey to the Moon – the hero being dragged by demons across the shadow of the Earth during an eclipse of the Moon. Although Newton had not yet formulated his theory of gravitation (indeed, he had not yet been born), Kepler's story contained reference to a point of balance between the attraction of the Earth and that of the Moon (rather like the neutral point). Kepler also made reference to the problems posed by the cold and lack of air in space.

In 1638, Bishop John Wilkins published a notable work, *The Discovery of a World in the Moone* in which he suggested that the Moon might be inhabited and that in the future the means might be developed to travel there and colonize it. In other tales of that century a voyage to the Moon was accomplished with the aid of a flock of giant swans, and attempts were made to travel by using solar energy to cause hot gas to escape from a chamber – the essence of the rocket principle.

The most remarkable story to emerge from the nineteenth century was Jules Verne's *From the Earth to the Moon* (1865) which, together with *A Trip Round the Moon* (1870), described a manned voyage round the Moon and back, and contained some remarkably accurate predictions. Verne's space capsule was fired from the 'Columbiad', a giant cannon set up in Florida, close to Cape Canaveral (site of America's major present-day launch complex), at a speed of 11 km/sec (40,000 km per hour) – the escape velocity. While it is unfortunately the case that any astronauts fired in this way would be squashed to pulp by the violent acceleration, and the capsule itself would be destroyed by atmospheric friction, in other respects the story was remarkably prescient. After approaching the Moon, but not landing, the projectile returned to Earth splashing down in the Atlantic Ocean where the travellers were picked up by a warship. The parallels with the Apollo 8 mission which took place about a hundred years later (see

Chapter 9) are remarkable – Verne got the launch site, velocity and mode of landing just about right.

By the late nineteenth century, serious scientific writers were laying the theoretical foundations for genuine spaceflight, and it was clear that the tool for the job had to be the rocket, the only means of propulsion able to operate in a vacuum.

The rocket works on the principle of reaction – Newton's third law of motion. By expelling material in one direction, the rocket itself is made to accelerate in the opposite direction, just as a bullet fired from a gun by an explosive charge causes the gun to recoil. If you were to stand on a sledge on a perfectly smooth sheet of ice, and fire a gun, you – together with the sledge – would begin to slide across the ice in the opposite direction to the bullet, and you would continue to move thereafter at a constant speed in a straight line (at least, until you came to the edge of the ice!). If you were to fire the gun again, you would pick up more speed, and if the sledge were equipped with a machine gun, you would continue to accelerate for as long as the gun continued to fire, each shot adding an increment to your speed. When the gun finally ceased to fire, you would continue to move at a constant speed thereafter.

The rocket is propelled by a stream of hot gas emerging from a narrow nozzle; each atom of that gas is a tiny 'bullet' fired off into space, and contributes its own minute 'kick' to the rocket vehicle. Another way of looking at the rocket principle is to think of a balloon. When the balloon's nozzle is sealed, the pressure of air inside the balloon is everywhere the same, and it will remain at rest. Open the nozzle and the pressure is no longer balanced; air rushes out in one direction (usually rather noisily) and the balloon flies off in the opposite direction.

The rocket is entirely self-contained. It needs no air in which to operate; indeed the presence of air is a hindrance, slowing the rocket's motion by friction. Other forms of propulsion require the presence of an external medium; the screw of a ship pushes against the water, the propeller of an aircraft pushes against the air. Even the jet engine (which uses the principle of reaction) draws in air to provide the oyxgen necessary to burn its fuel, and – of course – aircraft can fly only because they are supported by the lift generated by the airflow over their wings as they are propelled through the atmosphere. The rocket, on the other hand, is at its best in empty space.

Most large modern rockets – or 'launch vehicles' – derive their thrust from the chemical reaction between a fuel (such as liquid hydrogen) and an oxidant (such as liquid oxygen) which releases energy and expels a stream of hot exhaust gases; in other words, the oxidant provides the oxygen in which the fuel burns. In liquid-fuelled rockets fuel and oxidant are pumped from their separate tanks to a combustion chamber where ignition occurs. The term 'propellant' is used to describe fuel and oxidant taken together.

The final velocity attained by a rocket when it has used up all its fuel depends on two important factors– exhaust velocity and mass ratio. Exhaust velocity is the speed at which the hot gas is expelled from the rocket motor. The mass ratio is calculated by dividing the all-up weight of the rocket together with its payload (the 'cargo' to be carried into space) and a full load of propellant by the weight of the rocket plus payload when all the fuel has been consumed. The greater the mass ratio, the higher the final speed attained. If a rocket has a mass ratio of 2.72, if its launch weight is 2.72 times its empty weight, its final velocity will be exactly equal to its exhaust velocity. But here we come to a major technical problem; the escape velocity is 11 km/sec, but the highest exhaust velocities of liquid fuelled rockets are no more than 3 to 4 km/sec, so that to escape from Earth a rocket must attain a final velocity several times greater than its own exhaust velocity. This can be done, but to do so requires very high mass ratios. For example, to reach a final velocity of twice the exhaust velocity requires a mass ratio of 2.72 squared (2.72×2.72), i.e., 7.4, while to attain four times the exhaust velocity requires a mass ratio of $7.4 \times 7.4 = 55$, resulting in a requirement for a launch vehicle having 98 per cent of its launch weight made up of fuel and only 2 per cent available for its structure and payload! Considering the stresses to which a rocket is subjected in blasting off from the Earth, to build such a vehicle would seem to be impossible.

The solution adopted is the step rocket; the launch vehicle

is made up of a number of self-contained rockets stacked one on top of the other (or of a number of rockets clustered together round a central core rocket, known as the 'sustainer'). (Fig. 2.) Each stage carries it own propellant and its own rocket motor. As the first stage runs out of fuel, it drops away, and the second stage ignites to accelerate the remaining – lighter – part of the vehicle. When the second stage exhausts its fuel supply, it, too, drops away and the third stage motor takes over to produce further acceleration.

Even step rockets require exorbitant quantities of fuel to place quite modest payloads into space. For example, the giant Saturn V rocket used to take men to the Moon – and the most powerful launch vehicle yet to fly successfully – was capable of taking about fifty tonnes to orbit round the Moon, but weighed about 3000 tonnes on its launch pad, of which nearly 2,600 tonnes was propellant.

What is required is a propellant capable of yielding a much greater exhaust velocity, but there is little hope of any dramatic improvement in the chemical fuels currently in use; a completely different type of fuel is necessary to achieve a major improvement in the efficiency of rockets. Small-scale nuclear rockets – in which nuclear energy is used to heat fuel to higher temperatures and, therefore, to higher exhaust velocities – already exist; ion rockets, which employ electrical forces to accelerate ions – electrically charged atomic particles – to exhaust velocities ten to twenty times greater than those of chemical rockets, have also been tried out on a small scale. Neither of these types of rocket is capable of lifting itself off the ground, but they are capable of sustaining a low thrust for prolonged periods, so that used out in space they could sustain a gentle, but steady acceleration for sufficient time to achieve much higher final velocities. They cannot, however, take the place of chemical rockets for boosting payloads from the Earth's surface into orbit.

One of the great prospects for the future must be the fusion rocket which may, some day, be able to achieve exhaust velocities of around 10,000 km/sec. But that is very much a matter for the future (see Chapter 12); for the moment we are stuck with conventional rockets for some time to come.

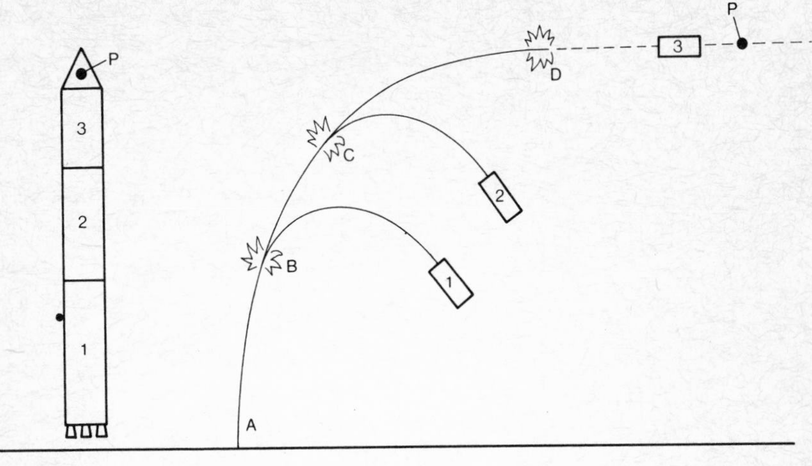

Figure 2.

(a) A three-stage rocket used to place a payload (P) in orbit. After lift-off (A) the first stage (1) accelerates the vehicle until first-stage burn-out (B) after which it drops away and the second stage (2) takes over to accelerate the vehicle to a higher velocity. After second-stage burn-out (C) the third stage (3) takes over to accelerate the payload to its required orbital velocity. The third-stage motor shuts down at (D) after which the payload continues in its orbit.

(b) An alternative to staging is the clustering concept. A number of rockets are clustered together round a central 'sustainer'. All the rocket motors fire at lift-off (left). When the boosters are exhausted (right) they drop off, leaving the sustainer to continue into orbit.

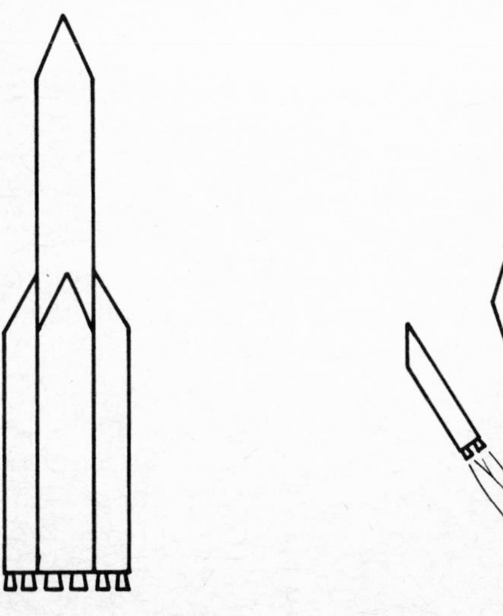

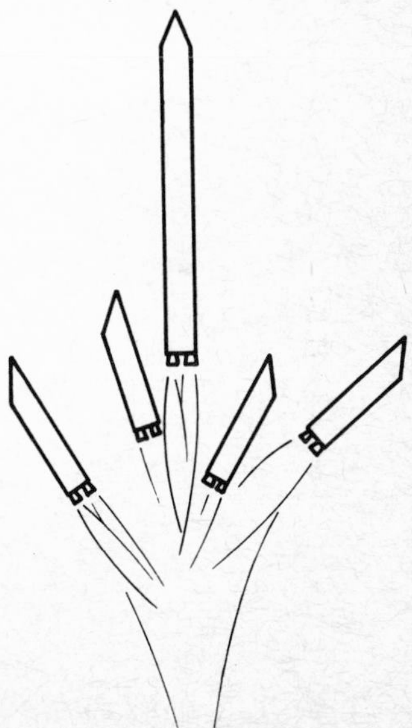

The history of the rocket

The invention of the rocket is usually ascribed to the Chinese, although the precise date is shrouded in the mists of time. The Chinese probably invented a form of gunpowder during the Han dynasty, about two thousand years ago, and it is well documented that batteries of rockets (or 'flaming arrows') were used by them to repel the Mongols at the town of Kai-fung-fu in the year 1232. There is reasonable evidence to suggest that rockets – attached to arrows – had been used as weapons of limited range and accuracy for several centuries prior to that date.

The first recorded device to employ the principle of reaction – key to the rocket – was an extraordinary and entertaining toy constructed around 360 B.C. by a Greek, Archytas of of Ternum. The device was a hollow chamber shaped like a bird and suspended from a stand; it was filled with water and heated, with the result that a jet of steam pouring from an aperture in the rear of the 'bird' propelled it in circles around the central stand. Again, in 62 A.D., Hero of Alexandria is said to have built an 'æolipile' – a spherical boiler, from opposite sides of which emerged two tubes emitting jets of steam; as a result the sphere spun round a central axis. Although neither of these gadgets had any practical application, they clearly demonstrated the principle upon which the rocket is founded.

By the thirteenth century knowledge of rockets had spread to Europe and from then on they were used from time to time in battles and sieges, albeit with limited effect. The basic idea of the step rocket was suggested as early as 1650 by Kasimir Simienowicz, a Pole who wrote of the possibility of attaining greater range with several rockets stacked one on top of another and igniting in turn. In 1687, of course, Newton laid down in his laws of motion the basic principles upon which the rocket is designed.

In 1792 and again in 1799 rockets were used against the British in India; these rockets, attached to bamboo canes, were effective over a range of about 1.5 kilometres. After this, the British began seriously to look into the rocket as a military weapon, and in 1804 William Congreve published a notable treatise, 'A concise Account of the Origin and

Progress of the Rocket System', showing the benefits of the rocket – for example, the lack of recoil – for firing from ships. He devoted considerable energy to improving the accuracy and performance of these somewhat erratic devices, eventually developing rockets which, fired from tubes, attained ranges in excess of 2 kilometres.

The first serious attempt to use Congreve's rockets in action ended in disaster when the fleet carrying rockets to bombard Boulogne was decimated by a gale in the English Channel, but the following year Boulogne was bombarded to considerable effect, and a similar major rocket attack was made on Copenhagen in 1807. The potential of the rocket – at least as a means of causing panic and confusion – was demonstrated in battle on a number of occasions during the next few decades, and the British rocket attack on Fort McHenry, Baltimore, in 1814 is commemorated by a line about 'the rocket's red glare' in the American national anthem.

Although further improvements were made during the nineteenth century – in particular, another Englishman, William Hale, developed a 'stickless' rocket using angled tail fins to spin the rocket round its long axis, greatly improving its stability in flight – it was overhauled in range and accuracy by conventional artillery and fell into disuse in the latter part of that century.

The next major theoretical steps were taken by the remarkable Russian schoolmaster and rocket pioneer, Konstantin Edouardovich Tsiolkovsky (1857-1935). The first of his many theoretical works, 'Free Space', was published in 1883; his most significant – written soon afterwards but not published until 1903 – was 'The Exploration of Cosmic Space by Reaction Vehicles'. Amongst his many achievements, he established the formula linking final velocity of a rocket to its exhaust velocity, and demonstrated that it would be necessary to use multi-stage rockets – staged either by stacking one on top of another, or by clustering rockets round a central core, the clustered rockets dropping away when their fuel was exhausted leaving the central 'sustainer' to accelerate further (the latter technique was adopted by the Soviet Union for its twentieth-century launch vehicles).

He also clearly demonstrated the advantages of liquid-fuelled rockets over solid-fuelled ones, and pointed out that the ideal fuel would be liquid hydrogen. The solid-fuelled rocket is a relative of the bonfire-night firework; it is basically a cylinder packed with combustible powder (e.g. 'gunpowder') which – once ignited – continues to burn until it is exhausted. Early solid-propellant rockets burned rather erratically and, since the 'combustion chamber' was the empty space inside the cylinder, the size of the chamber increased as the propellant was used up. The lack of control was one of the great drawbacks to the solid-fuelled rocket. For space applications the fact that the rocket cannot be switched off and restarted, and the fact that the whole structure has to be strong enough to contain the forces of combustion, are major drawbacks, too. (*Modern* solid-propellant rockets burn in a more controlled fashion and are effective devices used in a variety of shorter-range missiles and as the upper stages of a number of launch vehicles.) Liquid propellants are more efficient, more controllable, and are pumped from tanks into the combustion chamber of a rocket motor, which is the only part of the rocket vehicle which has to contain the heat and pressure of combustion; consequently the overall structure of the launch vehicle can be much lighter than it would be for a solid-propellant rocket.

The first giant leaps forward in practical rocketry were made in the United States. In 1919 Professor Robert H. Goddard published a highly significant paper entitled 'A Method of Reaching Extreme Altitudes' in which he pointed out the virtues of the rocket as a means of investigating the upper atmosphere, and made the suggestion that a rocket could be made to reach the Moon, marking its impact by means of flash powder. This suggestion made dramatic news in the *New York Times* under the headline 'Believes Rocket Can Reach Moon'.

Goddard worked on developing liquid-fuelled rockets and achieved the first successful launching of a rocket of this type on 16 March, 1926 at a remote farm in Massachusetts. Although the motor fired for a mere 2.5 seconds (and had to be ignited with the aid of a blowlamp), it attained a speed of just under 100 kilometres per hour and travelled a distance of about 60 metres. By 1935 his experimental rockets had attained altitudes of over 2 kilometres and velocities approaching 900 kilometres per hour; and in 1937 he experimented with rocket motors with steerable nozzles – forerunners of one modern method of steering rockets in flight.

Meantime, interest had been growing in Europe. Inspired by writers such as Hermann Oberth, whose 'Die Rakete zu den Planetenraumen' ('Rockets into Interplanetary Space') had been published in 1923, a group of German enthusiasts founded in 1927 the 'Verein fur Raumschiffart' (VfR) – the 'Society for Space Travel'); which within its first year grew to a membership of more than 500. Among its members was Wernher von Braun – later to become an architect of the American space programme. They immediately set to work to build rockets, and acquired some waste ground in the suburbs of Berlin on which to carry out test flights; this came to be known as the 'Raketen Flugplatz' (the 'rocket Flying Field').

The first successful flight of a liquid-fuelled rocket in Europe was achieved by Johannes Winkler on 14 March, 1931, and two months later the VfR fired its first 'repulsor' rocket to a height of more than 1 kilometre. By this time the activities of the VfR had attracted the interest of the German military, and in particular, of Colonel Dornberger, who headed the development of German rocket weapons prior to and during the Second World War.

Elsewhere, too, interest continued to grow. The American Rocket Society (as it eventually came to be known) was founded in 1930 and the British Interplanetary Society in 1933. Rockets were being developed in the Soviet Union at this time, one of the leading pioneers being Valentin Petrovich Glushko who, besides being deeply involved in the development of Soviet liquid-propellant motors, also made the suggestion that ion-electric rockets might be worth investigating. The first Soviet liquid-fuelled rocket to fly was developed by Sergei Korolev and Mikhail Tikhonravov and used a mixture of gasoline and liquid oxygen; on its first flight on 17 August, 1933 it reached a height of about 400 metres.

But it was in Germany between the mid-nineteen-thirties and mid-nineteen-forties that there took place the major steps in developing the rocket into a potent weapon with ranges of hundreds of kilometres. Dornberger established his rocket team, which included von Braun, at Kummersdorf West, south of Berlin, and, despite early setbacks, work continued in earnest to perfect the rocket as a serious military weapon. In 1934 a test rocket, known as the A-2, reached an altitude of 2.2 kilometres. As the work progressed, the need arose for a more secluded and extensive test range, and the operation was transferred to Peenemünde, an island in Stettin Bay on the Baltic coast.

By the end of 1937 the 6.7 metre-long A-3 vehicle had been built and had failed, but another class of test vehicle, the A-5, made its first successful flight in 1939. The long-range weapon itself was designated A-4 and was ready for test firings by 1942. After two disastrous failures the third A-4 lifted off on 3 October, 1942 and flew about 200 kilometres down range, landing within 4 kilometres of the aiming point.

Despite this success Hitler, in 1943, refused to increase the funding of the project for the extraordinary reason that he had *dreamed* that such weapons would never fall on England. Within a few months he had changed his mind, and during the later stages of the Second World War vast funds and some 20,000 men were poured into the project, in order to perfect and mass produce the A-4 which came to be known as the *Vergeltungswaffe-Zwei* (reprisal-weapon–2), the legendary V–2. The first operational V–2 was fired in July 1944, and by the end of the war in Europe over 4000 of them had been fired, including about 1500 directed at England many of which fell on London bringing with them sudden and unpredictable destruction.

The V–2 was 14.5 metres tall, with a maximum diameter of 1.65 metres, not including the span of the tail fins. Its launch weight was 12.9 tonnes of which nearly 9 tonnes was fuel. The propellant was ethyl alcohol and liquid oxygen, this giving an exhaust velocity of about 2 km/sec and a *thrust* (the force generated by the rocket motor) of 25 tonnes*.

* Strictly speaking the unit of force is the *Newton*, the force necessary to accelerate a mass of 1 kilogram at a rate of 1 metre/sec/sec. However, it is probably easier to get an impression of the magnitude of thrust developed by a rocket by thinking in terms of weight: a thrust of 1 tonne (1000 kilograms) is equal to the force necessary to support a weight of 1 tonne. A rocket weighing 1000 tonnes would require a thrust of at least 1000 tonnes to lift it off the ground. If the thrust generated were *exactly* 1000 tonnes, the rocket would simply hover on its launch pad until it had consumed enough fuel for the weight to drop below the value of the thrust; then the thrust would overcome the weight and the rocket would rise.

Steered in flight by moveable vanes in the exhaust, it had a maximum range of over 300 kilometres and reached heights of 80–95 kilometres at the peak of its trajectory. It carried near a tonne of high explosive. The altitude record attained by a wartime V–2 was 175 kilometres.

In 1945 von Braun and most senior members of his team surrendered to the American forces and they, together with about a hundred V–2s, were shipped to the United States where they continued to develop rockets for the U.S. Army team. The U.S. Army established a rocket test range at White Sands in the New Mexico desert in 1945, and in 1946 the first test firings of V–2s from American soil took place, the first one went spectacularly out of control, but on 17 December a V–2 reached a record altitude of 180 kilometres. Test firings of V–2s continued there until 1952, the ultimate altitude record being 214 kilometres.

A particularly significant event occurred in 1949. The Wac Corporal, a small 'sounding rocket' developed for studying the upper atmosphere, and itself capable of reaching heights of some 70 kilometres (with a boosted launch), was mounted on top of a V–2 to make the first serious two-stage rocket. One minute after lift-off on 24 February, 1949, the V–2 had reached an altitude of about 32 kilometres and a speed of about 1.6 km/sec; the Wac Corporal then ignited and fired for a mere 40 seconds, but in that time boosted its velocity to nearly 2.5 km/sec (about 9000 km per hour). The Wac corporal coasted up to a height of 403 kilometres eventually falling to earth about 130 kilometres down range. The principle of the step rocket – hinted at centuries before – had at last been proven, and a major step on the road to Space had been taken.

When Soviet forces occupied Peenemünde in 1945, they found that most key members of von Braun's team, the V–2s and most of the equipment had been spirited away to the United States. Only a handful of V–2s were located, together with parts, jigs and tools, and taken to the Soviet Union. The first firing of a V–2 from Soviet soil occurred on 30 October, 1947, from a site in Kazakhstan now known as the Kapustin Yar launch facility.

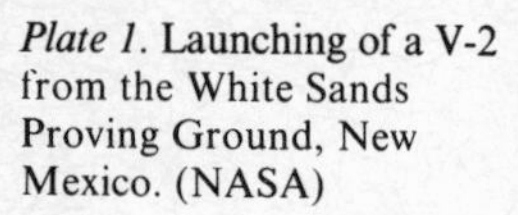

Plate 1. Launching of a V-2 from the White Sands Proving Ground, New Mexico. (NASA)

This was an uneasy time in world history. The world was polarized into the communist east and the capitalist west, each viewing the other with profound mistrust. On the face of it, during the decade following the Second World War, the United States had an overwhelming military superiority. They had the atom bomb and the means to deliver it at ranges of many thousands of kilometres with the aid of fleets of long-range bombers; in the West, at least, it was believed that the Soviet Union was technologically backward. Following the explosion of the first Soviet atom bomb in 1949, American President Harry Truman gave the go-ahead for the development of the even more devastating hydrogen bomb (the fusion bomb), which was first tested successfully in 1952. This lead was short-lived; the following year the Soviet Union, too, detonated a hydrogen bomb.

There emerged a dichotomy of approach to the means of delivering those awesome weapon. In the United States, which already had the means in the form of bombers, the attitude to long-range missiles was to hold back on developments until technology had reduced the size and weight of the bomb; in the Soviet Union work proceeded directly to build long-range rockets capable of carrying the original heavy weapon. In rough outline, the United States quartered the weight of the bomb and developed missiles to carry this lighter weapon, while the Soviet Union developed a much more potent rocket to carry the heavy bomb. As a result, when the urge to explore space asserted itself, the Russians had the means to launch far heavier payloads than the Americans.

With single-minded efficiency the Soviet Union had set about building powerful and effective long-range missiles. The SS–3, capable of carrying a nuclear warhead over a range of some 800 kilometres entered service in the early 1950s, but the first true intercontinental ballistic missile (ICBM), and key to the successes of the Soviet Space programme, was the SS–6 (known to NATO as 'Sapwood'). Built on the cluster principle, which had been suggested half a century earlier by Tsiolkovsky, it consisted of a central sustainer around which were clustered four liquid-propellant boosters. The sustainer and boosters each had four main combustion chambers, making twenty main engines, and in addition the sustainer had four smaller 'vernier' motors to control the altitude and direction of the vehicle; each of the boosters had two verniers, so that in total 32 rocket motors had to fire simultaneously when the vehicle took off. Standing some 28 metres tall on its launch pad, it was a rocket of staggering power compared to the V–2, and to anything existing in the U.S.A. at that time, the total thrust at lift-off being about 500 tonnes (compared to 25 tonnes for the V–2). After the vehicle blasts off – with all motors firing – the four boosters continue to fire for two minutes, and then fall away; the central sustainer continues to burn for a further 2.5 minutes thereafter.

Just two months after its first successful test firing (there were several earlier unsuccessful ones) this vehicle – given the

Plate 2. 'Business end' of a Vostok (A–1 series) launch vehicle of the type used to launch the first manned spacecraft. The four boosters – each with four main and two vernier motors – can be seen clustered round the central sustainer with its four main engines and four verniers. (Novosti Press Agency)

designation 'A' when used as a space launcher – hurled Sputnik 1 into orbit to the utter astonishment of the western world.

With an upper stage added it became the A–1, and was first used in this configuration to launch Luna–1 towards the Moon in 1959; the A–1 was capable of taking some 400 kilograms to the Moon or of placing nearly 5 tonnes in orbit round the Earth. The A–2, which has a further upper stage, can place 7.5 tonnes in close Earth orbit, and can take up to 1.2 tonnes towards Mars or Venus. Used for the Soyuz series of manned spacecraft (see Chapters 8 and 10), the A–2 together with its payload stands nearly 50 metres tall on its launch pad.

The most powerful Soviet launcher currently in use is the D–series, widely known as the *Proton* launch vehicle. The first stage comprises 6 boosters clustered round a central second stage which ignites after the boosters are exhausted; a third stage mounted on top of the second provides the final acceleration. First-stage thrust is about 1500 tonnes, and it is capable of lifting some 20 tonnes into a low orbit round the Earth. In its most advanced version it can send as much as 5 tonnes towards Mars or Venus. A very much more powerful launcher, the G–1, has been under development since the mid-nineteen-sixties. A prototype is believed to have blown up on its launch pad in 1969, and subsequent test launches in the following three years met with complete failure. It is believed that the rocket was originally conceived with the idea of a manned lunar landing in mind, but there is no official confirmation of this. However, it seems likely that the rocket will eventually make its appearance as a means of lifting heavy payloads into Earth orbit. Western estimates suggest the first-stage thrust will be between 4,500 and 5,000 tonnes, half as much again as the American Saturn V but, because of its using less energetic fuels, its payload to orbit capacity (estimates suggest about 150 tonnes) is unlikely to be very much more than that of Saturn V.

In the United States during the 1950s, developments in rocketry were fragmented, each of the services – Army, Navy, and Air Force having separate programmes to develop missiles for their own purposes. Von Braun's Army team developed the Redstone intermediate range missile – similar in capability to the Soviet SS–3; it owed much to its predecessor, the V–2, but had a thrust of 35 tonnes as against 25 tonnes. Von Braun's proposal, made in 1954 (see Chapter 1) to use the Redstone as the first stage of a vehicle capable of placing a satellite weighing a few kilograms into orbit was vetoed. However, a development of Redstone, the Jupiter–C – during a test of a re-entry nose cone – fired its payload to a height of 965 kilometres and across a range of 4,800 kilometres in September 1956, and there is little doubt that if he had not expressly been forbidden to do so, he and his team could, and would, have launched a small unofficial satellite prior to Sputnik 1.

Instead, of course, the U.S. Navy developed the Vanguard launch vehicle mentioned in Chapter 1. In its somewhat chequered career, which ended in September 1959, eleven launch attempts resulted in only three satellites placed in orbit – but a great deal of useful experience was gained.

In the meantime the U.S. Air Force was working on a highly sophisticated ICBM, the Atlas, which was of such lightweight construction that it had to be pressurized – rather like a balloon – to prevent its collapsing; the mass ratio was 25 – very high indeed. With a central core engine and two booster motors (all fed from the same propellant tanks) its thrust was about 150 tonnes compared to its lift-off weight of 118 tonnes. It was first used as a space launcher when the empty vehicle – known as Atlas–Score – was fired directly into orbit (with such a high mass ratio, this was possible without using an upper stage) on 12 December, 1958, carrying a tape recording of President Eisenhower's New Year greetings.

Used with the Agena rocket as second stage it could place about 2.5 tonnes into orbit, while with the more powerful Centaur second stage, it could place 8.6 tonnes in orbit, propel more than 1 tonne to the Moon and 0.6 tonnes towards Mars or Venus. The first firing of the Atlas-Centaur combination took the spacecraft Surveyor 1 to the Moon in 1966. The Atlas also played a key role in the American Mercury Project manned flights and in the Ranger, Surveyor, and Orbiter unmanned lunar missions, as well as the early Mariner series planetary missions. It is interesting to note that although the Atlas–Centaur had a lift-off thrust of less than one third of that of the Soviet A–1 vehicle, it could match its payload capacity purely as a result of greater propellant efficiency and superior mass ratio.

The United States also developed launch vehicles based upon the Thor intermediate range ballistic missile and the much more powerful Titan. The Titan is a potent rocket which has seen considerable development. For example, Titan II provided 195 tonnes of first-stage thrust and was used for the manned Gemini launches subsequent to 23 March,

1965. Later Titan III versions, launched with the aid of two strap-on solid-propellant boosters, achieve peak thrusts as high as 1300 tonnes, and can place over 13 tonnes in near-Earth orbit, take 3.8 tonnes to Venus or Mars, and up to 0.8 tonnes to the outer planets. Titan IIIE–Centaur combinations were used for the Viking and Voyager missions (see Chapter 5). Apart from the Space Shuttle, the Titan remains the most powerful American launch vehicle in service at present.

The largest American launch vehicle, and the most powerful the world has so far seen in operation, was the Saturn V developed for the Apollo lunar missions and described in Chapter 9; unlike the Atlas, Thor, and Titan, the Saturn series were developed as civilian projects with no direct military application. The Saturn V, with its payload, stood 110 metres tall, weighed nearly 3000 tonnes, and generated a maximum thrust of 3470 tonnes. The first one flew on 9 November, 1967, and the last to be fired, placed the space station 'Skylab' in orbit on 14 May, 1973. It was capable of placing 95 tonnes in orbit and of taking nearly 50 tonnes to the Moon.

Much more modest launchers, such as the Delta (originally based on the Thor IRBM) which has been in service in a variety of forms since 1960 and has, in its latest versions the ability to place about two tonnes in a low orbit, have provided a 'bread and butter' service for the launching of a wide variety of satellites. Of the 158 launches up to December 1981 only 6 have been outright failures. Another workhorse vehicle is the Scout, a four-stage solid-propellant vehicle used, also since 1960, to place lightweight satellites in orbit and to explore the upper atmosphere.

Although the first decades of the Space Age have tended to be regarded very much as a two-horse contest between the U.S.A. and the U.S.S.R., this really persisted only until 1965, for in that year France became the first nation outside of the two superpowers to develop an independent satellite launch vehicle. On 26 November of that year the French Diamant rocket placed a 38-kilogram satellite into orbit.

Next to join the exclusive club was Japan; after a number of failures in 1966 and 1967, the Japanese Lambda vehicle placed a small satellite in orbit on 11 February, 1970. Three months later the People's Republic of China launched a satellite which broadcast to a slightly astonished world the words of the song 'The East is Red'. The Japanese and Chinese have continued with their programmes, the Chinese in particular now having available an ICBM-based launcher able to place 2 tonnes in orbit.

On 28 October, 1971, the British Black Arrow vehicle placed Prospero, Britain's first and last independently-launched satellite into orbit; it is still in operation. Most recent of the nations to develop its own launch capability is India. On 18 July, 1980 an Indian designed and built four-stage solid-propellant vehicle (SLV–3) placed a 35 kilogram satellite into orbit.

The European Space Agency (ESA) which developed from the earlier European Space Research Organisation and the ashes of the ill-fated European Launcher Development Organisation (ELDO), now possesses a powerful and effective launch vehicle called Ariane which is able to compete effectively with the American Space Shuttle for the lucrative business of launching communications satellites. Ariane is described further in Chapter 11.

Despite the fact that these other nations have entered the field, it is still the case, in the third decade after Sputnik 1, that the space business is dominated by the two giants, the U.S.A. and the U.S.S.R.

These, then, are the principles which underlie getting into space, the rockets which first achieved it, and the current generation of launch vehicles which have made the launching of a satellite almost as mundane and commonplace as catching a Jumbo jet to fly the Atlantic. In the chapters which follow we shall examine some of the remarkable deeds accomplished by these launch vehicles and the satellites and spacecraft which they have carried across the space frontier.

4

Spaceprobes to the Moon

The Moon is our nearest celestial neighbour. Since the dawn of human history its silvery orb, lightening the blackness of night, marking the passage of time by repetitiously tracing out its monthly cycle of phases, has been an object of overwhelming interest and romantic fascination. Ever since Man first dreamed of flying to other worlds, the Moon has been the obvious first target.

The Moon travels round the Earth in a period of 27.3 days but, because of the Earth's motion round the Sun, and the fact that the phases of the Moon are determined by the relative positions of Sun, Moon, and Earth (see Fig. 3), the time interval between successive 'New Moons' is actually 29.5 days. This, the period of time during which it completes its cycle of phases, is termed the 'lunar month'.

The Moon rotates on its axis in precisely the same period of time as it takes to complete each orbit round the Earth. As a result of these two motions it keeps the same hemisphere permanently turned towards the Earth, and the far side of the Moon can never be seen from the surface of our planet.*

* In point of fact, due to phenomena known as 'librations' it is possible, over a period of time, for an Earth-based observer to see about 59 per cent of the total lunar surface. Because the Moon follows an elliptical orbit, its speed varies – being slower when it is further away and faster when it is closer in; the rate of axial spin, however, is constant. The two motions periodically get a little way out of step so that we can see alternately a little further round the east and round the west edges of the lunar disk. Likewise, because of the tilt of the lunar orbit and axis we can sometimes see a little way beyond the north and the south poles. 41 per cent of the surface remains permanently hidden.

However, it is important to note that the Moon does *not* keep the same face turned towards the Sun all the time; as it moves round the Earth, every part of the Moon, in turn, faces in the direction of the Sun. From the point of view of an observer on the lunar surface the time interval between two successive 'noons' amounts to 29.5 Earth days, the lunar 'day' and the lunar 'night' each lasting for about a fortnight.

The surface consists in part of lighter coloured highlands –

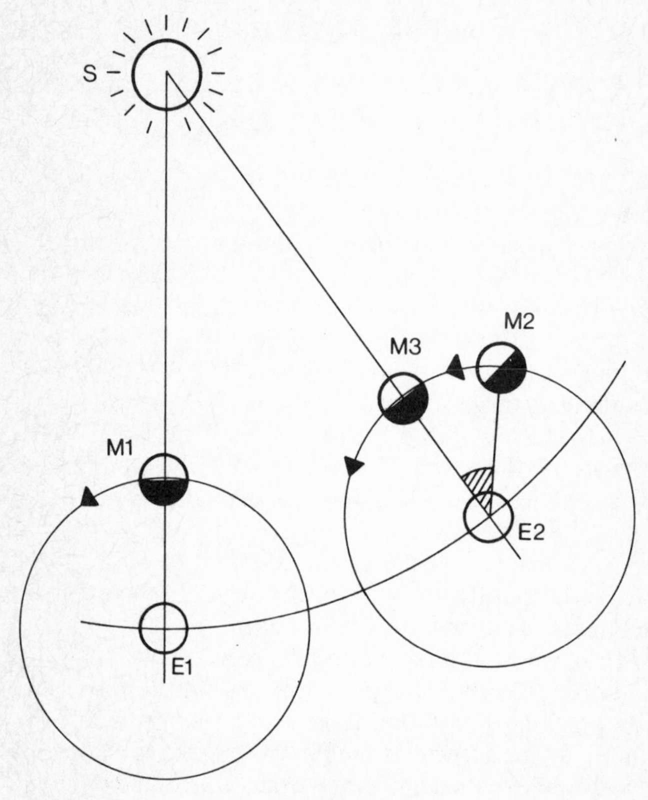

Figure 3.

Why the lunar month (the time between successive 'new moons') is longer than the orbital period of the Moon around the Earth. With the Earth at position E1 and the Moon at M1 it is New Moon and the dark side of the Moon faces the Earth. By the time the Moon has completed one circuit of the Earth it is at position M2. However, the Earth, in the meantime, has moved some distance along its orbit round the Sun. As a result the Moon must travel to position M3 before New Moon occurs once more.

mountainous terrain densely pock-marked with innumerable craters, ranging in size from giant formations such as Bailly, some 295 kilometres in diameter, down to tiny craterlets at the limits of visibility. On the Earth-facing hemisphere there are many dark, relatively smooth plains, which the early telescopic observers imagined to be seas and oceans, and named accordingly. The most conspicuous clearly defined 'sea', or 'mare' in Latin, is the Mare Imbrium in the Moon's north-eastern quadrant, a dark plain some 1300 kilometres across, bounded on three sides by high mountain ranges, and lying at an average depth below the mean lunar surface level of about 4 kilometres.

Of course, the term 'sea', is a misnomer. There is no water on the Moon, and no atmosphere. It is essentially a dead world (although vague isolated transient events have been observed, and gentle tremors – 'moonquakes' have been detected by instruments placed on its surface) with extremes of temperature ranging from nearly 120°C (about 390 K) at the lunar equator at noon, to about −160°C (about 110 K) in the middle of the lunar night.

Prior to the lunar probes there had been heated debate as to the process responsible for the formation of the craters; were they caused by the impact of giant meteorites, or were they the result of some kind of internal activity – volcanism? On Earth both types of crater are found, and it seemed reasonable to suppose that both types might be found on the Moon, too. The verbal battle between opposing factions concerned what was the *major* crater-forming process. The nature of the lunar 'seas' was also the subject of controversy. Were they relatively smooth lava plains upon which some future manned spacecraft could safely land; or were they – as had been suggested by astronomer Thomas Gold – covered with deep dust drifts into which a spacecraft would sink without trace? Earth-based observations could not resolve the issue. As is so often the case, where firm evidence was lacking, theories abounded.

In 1959 the dream of reaching the Moon was achieved, not by a man, but by a man-made object, the Soviet spacecraft Luna–2 which, at 21 hours 2 minutes 23 seconds G.M.T. on 13 September, plunged on to the lunar surface at a speed of nearly 12,000 kilometres per hour after a 34-hour flight from

Earth. The impact of the 1.5 tonne spacecraft occurred about 435 kilometres from the centre of the visible disk, not far from the prominent crater Archimedes in the Mare Imbrium ('Sea of Rains').

Luna–2 had been preceded some nine months earlier by Luna–1 which is generally believed to have been an attempt at a crash landing, but which missed the Moon by the small margin of between 5 and 6 thousand kilometres. The spacecraft, with a total weight of 1472 kilograms (including 361 kg of instrumentation) was launched from Tyuratam on 2 January, 1959 and made its closest approach to the Moon two days later. Like Luna–2, it was first fired into a parking orbit round the Earth, then its motor was fired again to accelerate the craft beyond escape velocity and place it on a trajectory towards the Moon. This is no easy task: the motion of the Moon has to be allowed for; in two days the Moon travels about 175,000 kilometres along its orbit, so the spacecraft must be aimed towards the point where the Moon *will be* when the spacecraft is due to arrive. Luna–1 was the first spacecraft to be fired from a parking orbit, and the first to achieve escape velocity.

The Russians gave the spacecraft the name 'Mechta', which means 'Dream'. This seems singularly appropriate since to reach the Moon had been a long-standing dream of mankind, and Kepler in the seventeenth century had written a story of that title, about a mystical voyage to the Moon. In the West, the name 'Lunik' was attached, but 'Luna–1' is now more generally used to be consistent with the rest of the series.

After passing by the Moon, Luna–1 entered an orbit round the Sun, and in so doing became the first artificial planet.

The flight of Luna–1 was by no means the first attempt to reach the Moon. During the previous year the United States made no less than four unsuccessful attempts to reach our natural satellite. The first attempt took place on 17 August, 1958, when a 38-kilogram spacecraft (now designated 'Pioneer 0') was fired by the U.S. Air Force on board a Thor-Able 1 vehicle: 77 seconds into the mission, the lower stage exploded thereby terminating the flight rather abruptly.

Two months later on 11 October, the first deep space mission undertaken by the American National Aeronautics and Space Administration (NASA) – the civilian space agency which had officially come into being just ten days previously – very nearly succeeded. When the motor shut down the final velocity was a mere 2 per cent short of the required value – insufficient to allow the spacecraft to escape from the Earth's clutches, but enough to take Pioneer 1 to a height of 113,800 kilometres before falling back and breaking up over the Pacific Ocean the following day. The following month Pioneer 2 fell back to Earth when the third stage of the launcher failed to ignite. Pioneer 3, launched on 6 December by the U.S. Army team's Juno II vehicle failed to reach escape velocity, but nevertheless rose nearly as high as Pioneer 1 before plunging back to Earth. It discovered the outer Van Allen radiation zone, so the mission was not in vain.

The new year had barely dawned when Luna–1 made its dramatic close encounter with the Moon. The Americans doggedly attempted to catch up. Pioneer 4, launched on 4 March, 1959, missed the Moon by some 60,000 kilometres and sped onwards into interplanetary space to become, like Luna–1, an artificial planet. Communication was maintained with the spacecraft to a range of about 660,000 km, a record at that time.

Two more Pioneers failed spectacularly before 1959 was out, but the Soviet Union crowned its 'year of the Moon' with another sensational 'first' – pictures of the lunar far side, hitherto unseen by human eye. On 4 October, exactly two years after the launching of Sputnik 1, Luna–3 soared skywards from Tyuratam with an instrumental package of some 435 kilograms aboard. Luna–3 followed a long elliptical trajectory which took it to apogee at a distance of 470,000 kilometres on 10 October. By 7 October, however, the spacecraft had passed by the Moon and lay about 60,000 kilometres beyond Earth's satellite. At that time the Moon was a thin crescent as seen from Earth; consequently most of the far side was illuminated by the Sun. The cameras were switched on for a 40-minute photographic session, but we on Earth had to wait for some time to see the results. The film was developed on board and the pictures broadcast to Earth when Luna–3 came to perigee on 18 October, and was then at a distance of just under 47,000 kilometres; the power required for their transmission was then at a minimum. The pictures were scanned on board by a television system, broadcast, and reconstructed on Earth.

Although the pictures were of lamentably poor quality compared to those which can now be received from the depths of the Solar System, and showed no more detail than can be seen on the Earth-facing hemisphere with a modest pair of binoculars, they were of the utmost interest none the less, and represented a major technological achievement. Not quite all of the far side was shown, for it was essential to include part of the visible hemisphere in order to orientate the photographs and correlate far-side features with near-side ones; and in any case, not all of the far side was illuminated at the time. The most obvious near-side feature shown in the Luna–3 pictures was the compact Mare Crisium (Sea of Crises), familiar to all lunar observers.

In some respects, because of their poor quality, the photographs led to errors of interpretation; thus a bright streak, tentatively named the 'Sovietsky Mountains', turned out eventually to be no more than it seemed – a bright streak across the lunar surface. A deep, dark-floored crater, which was positively identified on these early pictures, was named 'Tsiolkovsky' in honour of the great Russian rocket pioneer. What was very clear, however, is that the far side of the Moon is strikingly different from the near side in one crucial respect – the virtual absence of dark mare regions. Large depressed areas are present, but in most cases they are not filled with the dark basaltic material which characterizes the familiar 'seas'. It is now believed that the crust on the far side is appreciably thicker than on the near side, and that this has largely prevented the darker material from the lunar interior from flooding out to fill the basin floors. Evidently the tidal forces exerted by the Earth on the Moon have played a role in bringing about this situation. (Tidal forces were also responsible for slowing down the Moon's rate of rotation on its axis until the lunar day was equal in duration to its orbital period round the Earth.)

1959, then, was a remarkably successful year for the Soviet space programme which achieved the first close fly-by (Luna–1), the first 'hard' landing (Luna–2) and the first

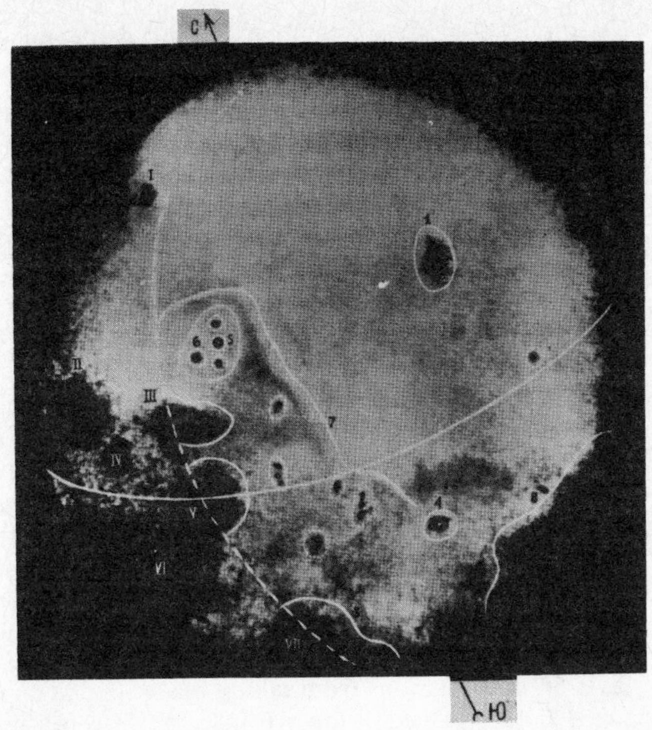

View of the far side of the Moon as photographed in 1959 by Luna–3, the first spacecraft to photograph the Moon's hidden hemisphere. Near-side features on the left are indicated by Roman numerals; for example, feature II is the Mare Crisium (Sea of Crises). (Novosti Press Agency)

pictures of the far side (Luna–3). The next few years came as an anticlimax so far as lunar exploration was concerned: 1960 was notable only for two further Pioneer failures: 1961 produced the first steps in the American *Ranger* programme, the aim of the Ranger spacecraft being to transmit photographs of the lunar surface right up to the moment of impact. Rangers 1 and 2 – test vehicles which were not intended specifically to go to the Moon, failed to achieve escape velocity, and must be counted as failures. Ranger 3, launched on 26 January, 1962, made a more promising start, but nevertheless – due in part to the Atlas–Agena launch vehicle attaining *too high* a final velocity – missed the Moon by some 37,000 kilometres.

The Ranger spacecraft was a sophisticated one, carrying a

range of instruments to measure quantities such as solar radiation and magnetic fields, in addition to having the capacity to transmit thousands of television pictures during the approach to the Moon. In addition, Rangers 3 to 5 inclusive also carried a small, elaborately cushioned instrumental package which was intended to survive the impact and transmit data directly from the lunar surface.

Ranger 4, launched on 23 April, 1962, became the first American spacecraft to reach the Moon but, unfortunately, effective control of the spacecraft was lost soon after launching, and it crashed on to the lunar far side on 26 April – the first spacecraft to achieve that dubious distinction. Ranger 5, in October, missed the Moon by a mere 725 kilometres, but a power loss ensured that no pictures were received. No further Ranger launchings took place until 1964. In 1963, the first named Soviet lunar probe for more than three years – Luna–4 – missed the Moon by a wide margin.

The troubled Ranger programme resumed in January 1964. Ranger 6 struck the western edge of the Mare Tranquillitatis (Sea of Tranquillity) a mere 32 kilometres from its nominal aiming point, but a short circuit in its power supply prevented any television transmissions from taking place; thus a minor component failure turned Ranger 6 into the fifteenth consecutive failure in the accident-prone American lunar programme. Few observers, in the face of this catalogue of disaster, would have believed that five and a half years later, men would set foot on the surface of that same lunar sea.

NASA's persistence ultimately was rewarded on 31 January, 1965, when Ranger 7 plunged into the Mare Nubium (Sea of Clouds) within 15 kilometres of its aiming point. The cameras were switched on when the craft was some 2000 kilometres above the Moon, and as Ranger 7 accelerated to its impact velocity of over 8,000 kilometres per hour, a total of 4,316 television pictures were returned. The last frame covered an area measuring only about 30 metres by about 45 metres, and clearly revealed features (including tiny crater-like pits) only tens of centimetres across, thus revealing details about two thousand times finer than could be photographed by Earth-based telescopes. Success, when it finally came, was sweet indeed.

Rangers 8 and 9 rounded off the series in highly successful fashion, the former returning 7,137 pictures and the latter, 5,814. Ranger 9 hit the floor of the crater Alphoneus, within 9 kilometres of the aiming point on 24 March, 1965. The pictures were displayed live on American television as the spacecraft hurtled to destruction, so allowing American viewers to see for themselves the lunar surface rushing up to meet them. High drama indeed!

In 1965 there was too a renewed burst of Soviet lunar activity, but with little success. Lunas 5, 6, 7, and 8 all failed to achieve their objectives – undoubtedly they were intended to soft-land on the lunar surface – mostly due to failure of the retro-rocket designed to break the spacecraft's fall. Zond 3, an interplanetary spacecraft with the astonishing weight of 8.9 tonnes, passed by the Moon while en route towards the orbit of Mars, and took 25 photographs of the far side of much better quality than those taken by Luna–3. Processed on board, then scanned, these pictures were repeatedly transmitted to Earth as a test of the communications sytem, until the craft reached a range of 3,500,000 kilometres.

The following year saw at least 10 spacecraft (five American and five Russian) attempting a wide range of exploratory missions. Of these missions the most dramatically successful was Luna–9, launched from Tyuratam on 31 January, 1966. Three days later, as the craft was less than 100 kilometres above the lunar surface, and falling at a speed of about 8,000 kilometres per hour, the retro-rocket was fired to brake its descent. The instrument package, weighing 100 kilograms, was ejected to make a rather hard 'soft-landing' in the Oceanus Procellarum (Ocean of Storms), the large dark plain which occupies much of the visible face of the Moon at last quarter. The egg-shaped instrument module opened out its protective 'petals', and the cameras went into action.

On 3 February, for the first time ever, a man-made object had landed safely on another world.

The camera lenses were mounted at a height of only 0.6 metres from the ground and, taking account of the fact that the Moon is much more sharply curved than the Earth, the horizon to which the cameras could 'see' was decidedly limited. Nevertheless 27 photographs were sent back before the module's batteries gave out, and these revealed surface details as small as a few millimetres in diameter. Most dramatic of them all was the first picture which showed a small rock (about 25 cm in diameter) close to the spacecraft, casting a long shadow because of the low altitude of the Sun at the landing site. Perhaps the most important result of all, which was immediately apparent, was that the surface had sufficient bearing strength to support the spacecraft; Luna–9 had not sunk out of sight into the lunar dust – the 'dust drift' theory was disproved at a stroke, and we could be confident that future manned spacecraft would not sink irrevocably from view.

Exactly two months later, Luna–10 became the first artificial lunar satellite. A spacecraft from Earth, approaching the Moon, and left to its own devices, would do one of three things: collide with the Moon, loop round the far side and return to the vicinity of the Earth, or carry on into interplanetary space. To place a satellite in orbit round the Moon, it is necessary to fire a rocket motor to slow down the craft to a speed of less than the lunar escape velocity. In the case of Luna–10, the motor was fired as the spacecraft was looping round the far side, and as a result it entered an elliptical orbit ranging in height above the lunar surface from 350 kilometres to 1017 kilometres. Although no photographs were returned, the satellite showed that the lunar magnetic field was extremely weak and made measurements of gamma radiation from the surface.

Within a further two months the Americans had matched and far surpassed the achievement of Luna–9 with the first of their Surveyor series. The spacecraft – fired from Cape Canaveral on the first flight of the new Atlas-Centaur launch vehicle – touched down in the Oceanus Procellarum (about 800 kilometres from Luna–9) at the gentle speed of 13 kilometres per hour on 2 June, 1966. Surveyor was powered by solar panels and, as a result, had a far longer operational lifetime than Luna–9. The cameras were mounted higher (1.6 metres above the surface) and this increased its range of vision compared to the Soviet craft. The landing had taken place two days after local sunrise, and photographic activity continued to the end of that lunar day, on 14 June, by which time 10,338 pictures had been taken under varying angles of illumination. Measurements were also made of the bearing strength of the lunar surface, and of its physical nature.

The spacecraft survived the lunar night and was switched on again on 6 July, and during the second lunar day a further 899 pictures were taken.

After the long succession of failures which marked the early stages of the Ranger programme, the instant success of Surveyor came as a joyous event for the NASA staff and for the whole American space programme. The fact that Surveyor 2 crashed uncontrollably on 23 September, 1966, was accepted quite philosophically! Three out of four Surveyor missions were highly successful in 1967 (these were Surveyors, 3, 5, and 6; Surveyor 4 crashed), and the series was brought to a conclusion in January 1968 by Surveyor 7, the last *unmanned* American spacecraft to land on the Moon.

Among the highlights of the Surveyor programme, mention must be made of Surveyor 3 which landed on 20 April, 1967, some 370 kilometres south of the crater Copernicus. As a result of a failure in the landing radar system the fine control thrusters did not cut out when the spacecraft reached the surface with the result that Surveyor 3 bounced twice and landed three times before the motors finally switched off! None the worse for its experience the spacecraft functioned perfectly. For the first time a mechanical scoop was used to dig trenches in the lunar soil and to pick up lunar pebbles, allowing a detailed analysis of lunar material. Among the 6,315 pictures taken were some of the Earth in the lunar sky, and of a magnificent total eclipse of the Sun by the Earth which took place on 24 April.

Contact was lost at lunar nightfall on 3 May, but this was not the last to be heard of Surveyor 3. Two and a half years later, in a feat of incredible navigational precision, the lunar module 'Intrepid' (of the Apollo 12 mission) touched down just 180 metres from the dormant spacecraft, and astronauts Conrad and Bean were able to walk across to retrieve parts of the craft for return to Earth (see Chapter 9). The first interplanetary scrap merchants had arrived!

By analysing alpha particles (particles released from a

radioactive source on board the craft) scattered back from the lunar surface an analysis of the chemical composition of the lunar rock could be made, and in this way, Surveyor 5 showed that lunar rock was similar in nature to terrestrial basalt – a volcanic rock typical of the ocean beds. Surveyor 6 sent back a record 30,065 pictures, and Surveyor 7, which landed on the outer slopes of the conspicuous 80-kilometre diameter crater, Tycho, was able to sample the ejecta flung out by the impact which produced this – one of the youngest major craters – perhaps about 3 billion years ago.

With the forthcoming Apollo manned landing missions very much in mind, NASA embarked upon a programme of detailed surface mapping with the first of the Orbiter series spacecraft in August 1966. For this type of mission direct television transmission in 'real time' was not required (on the Ranger missions, pictures had to be transmitted immediately, for obvious reasons!); instead, the image was formed on a photographic film, then scanned electronically and broadcast back to Earth. The whole process was decidedly leisurely, for it took more than 40 minutes to scan and transmit each picture. The 60-metre roll of film carried on board was sufficient to take more than 200 'frames' where each frame consisted of one medium resolution (wider angle) and one high resolution (narrow angle) picture. Orbiter 1 sent back a total of 414 pictures including views of nine 'primary' and seven 'potential' Apollo landing sites, but after its triumph, met a rather ignominious end when it was deliberately crashed to make way for Orbiter 2 which entered lunar orbit in November, 1966.

Most dramatic of Orbiter 2's pictures was an oblique-angle frame of the giant crater Copernicus which showed, in breathtaking relief, the terraced walls and central mountain peaks of this majestic formation. Hailed at that time as 'The Picture of the Century' (a claim which has been surpassed many times since by other spectacular photographs) it aroused wide popular interest and was carried boldly in many national newspapers.

Orbiters 3 to 5 completed the mapping project in 1967, covering the whole lunar surface at high resolution, apart from small regions close to the poles. By then it could be said

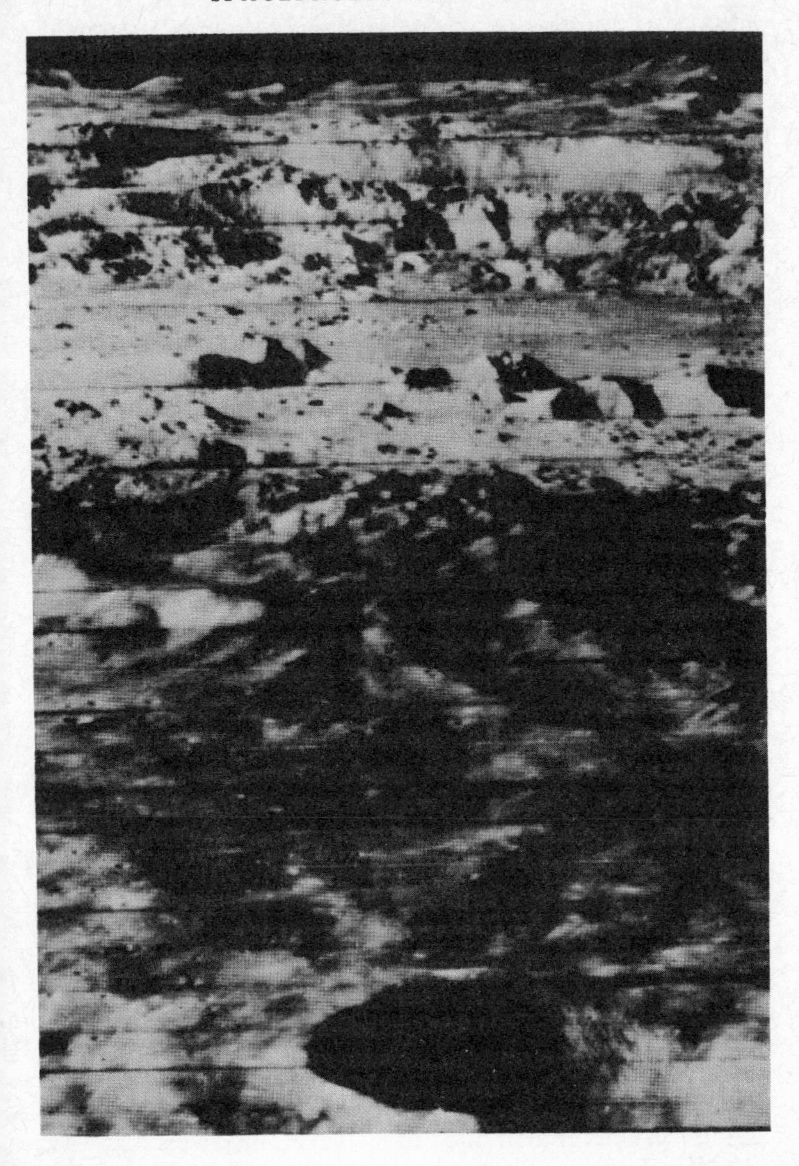

Plate 4. The lunar crater Copernicus photographed from an oblique angle in 1966 by Lunar Orbiter 2. When it first appeared, this photograph was hailed as 'The Picture of the Century'. Although many more spectacular pictures have been seen since that time, this one still gives a vivid impression of the terraced walls of this mighty crater. (NASA)

with justification that the Moon had been mapped to greater, more uniform, precision than the Earth itself. Another fundamental discovery made by the Orbiters was the existence of mascons (*mass con*centrations), concentrations of dense matter under certain regions of the surface such as the Mare Crisium and Mare Imbrium. These concentrated masses – believed to be made up of lava in and below the mare regions – caused deviations in the motion of the Orbiters as they passed over them.

By the end of the Orbiter and Surveyor programmes, the American space programme was concentrated almost entirely on the goal of landing men on the Moon (see Chapter 9). Since Surveyor 7, no *unmanned* American spacecraft has landed on the Moon.

Following on the successes of Lunas 9 and 10, the Soviet Union continued its assault on the Moon in 1966 with three more successful missions, Lunas 11 and 12, which were orbiters, and Luna–13 which touched down inside a shallow crater on the Oceanus Procellarum on Christmas Eve. For the first time a mechanical arm was able to drill down into the lunar soil to measure its strength and temperature. The Americans had the Moon to themselves in 1967, but the following year, Soviet activity picked up once more. In April, Luna–14 went into orbit, and later that year two very heavy spacecraft (Zond 5 and Zond 6) – which are thought to have been capable of carrying a human crew – looped round the Moon and returned to Earth (see Chapter 9), the first unmanned spacecraft to achieve this feat. Similar flights were made by Zonds 7 and 8 in the following two years.

Luna–15 was placed into orbit round the Moon in July 1969, just a few days before the manned Apollo 11 spacecraft arrived, preparatory to carrying out the first manned landing. It is widely believed that the aim of the mission was to land, collect a sample of lunar material, and return it automatically to Earth before the Americans could do so, and without having 'risked human life' to achieve this. Certainly the propoganda value of such a coup – had it been achieved – would have been considerable, but the suggestion remains no more than a very plausible speculation. In the event, Luna–15 began its descent but crashed into the Mare Crisium on 21 July, the very day on which Neil Armstrong and Edwin Aldrin became the first men to set foot on the Moon. That the impact should have occurred in the 'Sea of Crises' seems strangely appropriate.

More than a year was to pass before that aim was realized. On 20 September, 1970, Luna-16 touched down on its four stubby legs just south of the lunar equator in the Mare Foecunditatis (Sea of Fertility) some 900 kilometres distant from 'Tranquillity Base', the Apollo 11 landing site. On command from the ground-based controllers, Luna–16 deployed a mechanism to drill a core sample to a depth of about 15 centimetres, then to withdraw it carefully and retract it into the upper stage. On 21 September the upper stage, with its precious cargo aboard, blasted off the now useless descent stage of the vehicle, back home to Earth. Three days later the re-entry capsule – charred by the searing heat of re-entry, but undamaged – parachuted down in Kazakhstan with its 100-gram sample of lunar material intact. The first ever 'remote sample return' mission had been accomplished.

Analysis of the sample by the Soviet Academy of Sciences, showed that the material from this site was broadly similar in composition to that obtained from the Apollo 11 mission; there were some detailed differences, but in essence the principal minerals identified in lunar rocks proved to be the same as those found as the major constituents of terrestrial rocks.

Within two months the Soviet space programme achieved another coup. When Luna–17 touched down gently on the Mare Imbrium on 17 November, ramps were lowered to the surface and down these ramps rolled the most improbable-looking and ungainly vehicle imaginable: Lunokhod 1. The vehicle looked like a mobile bath tub, the 'bath tub' itself, some 2.2 metres in diameter, acted as an instrument container, the lid of the bath tub being opened when the vehicle was operating. The underside of the lid (which was turned towards the Sun when the lid was open) was covered with solar cells to provide electrical power for the vehicle. During the cold lunar night the lid was sealed to protect the instruments, and battery power maintained a tolerable temperature within the

Plate 5. Lunokhod–1, the first remote-controlled lunar roving vehicle. (Novosti Press Agency)

instrument container. The bath tub was mounted on eight independently-driven wheels, and forward facing television cameras relayed the view ahead to Earth-based 'drivers' who could steer the vehicle round obstacles. Four steerable television cameras gave higher-quality panoramic views of the lunar surface all around the vehicle, while the surface was probed by means of an x-ray spectrometer and a 'penetrometer' (to measure strength). Measurements were also made of cosmic rays and cosmic x-rays. In addition the vehicle carried a French-built laser reflector for measuring the distance from Earth to Moon with great precision.

The vehicle pottered along, stopping to take measurements every few metres. As the Sun set on 27 November, the vehicle was parked and closed up for the long lunar night. On 9 December a signal went out from the controllers in the Crimea and – to the delight of all concerned – Lunokhod 'awoke' from its slumbers, and resumed operations. During its operational lifetime, which lasted for 11 months, it travelled 10.5 km, covered a total area of some 80,000 square metres, returned over 20,000 pictures and made surface measurements at hundreds of sites. The Earth-based team controlled the vehicle with a high degree of success, avoiding obstacles and extricating Lunokhod from tricky situations, including at least one occasion when it was very nearly bogged down in soft soil. Despite its somewhat bizarre appearance it proved to be most successful, and clearly demonstrated the potential of remote-controlled vehicles for surface exploration.

Lunokhod 2 was delivered by Luna–21 to the Mare Serenitatis (Sea of Serenity) in 1973, and travelled considerably further than its predecessor, covering 37 kilometres and returning over 80,000 pictures. However, it ceased to function after four months, possibly as a result of some mishap which befell it in its lonely wanderings.

Of the remainder of the Luna programme, Lunas 18 and 23 crashed, Lunas 19 and 22 became successful lunar satellites, and Luna–24 touched down gently in the Mare Crisium on 18 August, 1976. After drilling out a sample to a depth of two metres (more than ten times deeper than Luna–16's drilling) the upper stage successfully returned its sample to Earth on 22 August; the samples turned out to be of the highest scientific interest. Since that time no spacecraft, manned or unmanned, has been to the Moon.

During the eighteen years of lunar exploration, at least in the years prior to the Apollo landings, the Soviet and American programmes were carried out in competition. So far as unmanned missions are concerned, most of the major 'firsts' were achieved by the Russians (first close fly-by, first hard landing, first view of the far side, first soft, or soft-ish, landing, first lunar satellite, first remote sample return, and first remote-controlled roving vehicle), but after many early failures, the later Rangers, the Surveyors, and the Orbiters showed the superiority of American photographic and data-transmission techniques. The Apollo lunar landings, of course, crowned mankind's assault on the Moon, but that story is covered in Chapter 9.

In its breadth and scale, the exploration of the Moon, within an eighteen-year period, surpassed the dreams of all but the most ardent of the early visionaries. As a result, our understanding of the Moon has been immeasurably enhanced – but many problems and puzzles remain, and as with all scientific investigations, new problems have been thrown up to tax our ingenuity.

Despite the bombardment which it has suffered at the hands of mankind, the romance of the Moon has not been shattered. Rather, it has been enhanced.

5

Robot Explorers of the Solar System

Such was the pace of events in the first few years of the Space Age that the exploration of the Moon had barely begun when the first attempts were made to reach the planets. It is generally believed that the Soviet Union set out to launch probes towards Mars in 1960, but the spacecraft failed to reach their required parking orbits round the Earth, and were unable to launch forth into interplanetary space.

The first spacecraft to make a close encounter with another planet was Venera 1, a Soviet spacecraft launched towards Venus on 12 February, 1961. Radio contact was lost after fifteen days, by which time the spacecraft was at a range of 7,500,000 kilometres, but there is every reason to believe that it passed within 100,000 kilometres of the planet on 19 May. The first successful fly-by of Venus was achieved by the American spacecraft Mariner 2, which by-passed Venus at a range of 34,850 kilometres on 27 August, 1962, after a journey lasting 109 days. A foretaste of the revolution in our knowledge of the planets which the spacecraft were to bring in the next two decades was given by the results obtained; Mariner 2 showed, among other things, that Venus had a temperature far in excess of what astronomers had believed up to that time.

Before looking more closely at the achievements of the planetary spacecraft over the past twenty years, it is useful to think about the orbits they pursue and the speeds they must achieve.

In Chapter 3 we saw that to reach interplanetary space, a spacecraft must exceed the Earth's escape velocity (approxmately 11 km/sec). If a spacecraft is launched with only this minimum velocity, by the time it has travelled about a million kilometres from the Earth, its velocity (relative to the Earth) will have dwindled almost to zero, and it will not be able to reach any of the planets; instead it will enter an orbit round the Sun, similar to the orbit of the Earth. This was the fate which befell some of the early lunar probes which missed the Moon. A spacecraft bound for the planets must be launched at a speed in excess of escape velocity, but as Table 2 (page 74) shows, a modest increase in launch velocity can achieve a considerable gain in final velocity.

Once the spacecraft has 'escaped' from the Earth, its trajectory is governed by the gravitational field of the Sun. Depending upon its velocity, the spacecraft will follow a circular, elliptical, parabolic, or hyperbolic path relative to the Sun, and will behave just like a tiny planet, asteroid, or comet. The average speed of the Earth in its orbit is just under 30 km/sec, this being the value of circular velocity, relative to the Sun, at a distance from the Sun of 1 astronomical unit. A spacecraft moving more slowly than the Earth will follow an ellipse which takes it closer in towards the Sun, whereas a spacecraft moving faster than the Earth will follow a trajectory that takes it further out into the depths of the Solar System.

A spacecraft moving faster than 42 km/sec will escape from the Solar System altogether, this being the escape velocity of the Sun at a distance of 1 astronomical unit. Strangely enough, it is considerably easier to escape from the Solar System than to reach the Sun! A spacecraft launched from Earth at the Earth's escape velocity will end up moving round the Sun at the same speed as the Earth – 30 km/sec, and need attain only an additional 12 km/sec to reach Solar System escape velocity. To reach the Sun, however, it must cancel out the initial motion imparted by its launch platform (the Earth); after escaping from Earth it requires an additional 30 km/sec, in the opposite direction to the Earth's motion, to cancel out the 'sideways' motion and allow it to drop directly – like a stone released from an enormous height – into the Sun. If the sideways motion is not completely cancelled, the spacecraft will follow a narrow ellipse which skims past the solar surface and back out towards the orbit of the Earth.

Although it might appear that the simplest way to reach a planet is to wait until it (Mars, or Venus, say) was at its closest, and fire the craft straight across the gap, so taking the shortest possible route, in fact no rocket yet built is capable of achieving this. To achieve this the launch vehicle would not only need to supply the craft with enough energy to 'climb' directly away from the Sun to Mars, say, but it would also have to cancel out the 'sideways' motion of the Earth which, as we have seen, amounts to 30 km/sec (i.e. about 108,000 km/hour). Apart from being impossible at present, such a mission would be extremely wasteful of fuel. It is far better to launch the craft in such a way as to use the Earth's motion to the best possible advantage. The most economical routes to the planets were worked out in 1925 by the German town planner and architect, Walter Hohmann, in his publication *Die Erreichbarkeit der Himmelskorpen* (The Attainability of Heavenly Bodies).

Hohmann argued that the least amount of energy would be required if the spacecraft were launched in the same direction as the Earth is moving. Relative to the Sun, the spacecraft would be moving, after 'escape' from the Earth, with a velocity equal to its 'final velocity after escape' plus the velocity of the Earth itself. If the velocity were carefully judged, the spacecraft would follow an ellipse (see Fig 4) which just reached out to the orbit of the target; the perihelion distance would be equal to the radius of the Earth's orbit, and the aphelion distance would be equal to the radius of the target's orbit. To reach an inner planet (Mercury or Venus) the spacecraft must be fired in the opposite direction to the Earth's motion so that its velocity *subtracts* from that of the Earth, and it enters an ellipse which reaches closer to the Sun, meeting the target at perihelion.

A trajectory which takes a spacecraft from one orbit to another is called a *transfer orbit*; a minimum energy transfer ellipse is called a *Hohmann transfer orbit*. Although it is the most economical way to go, the penalty to be paid for following such an orbit, is the long flight time involved compared to that required for shorter, but more expensive, 'fast' trajectories. To reach its target via a Hohmann transfer

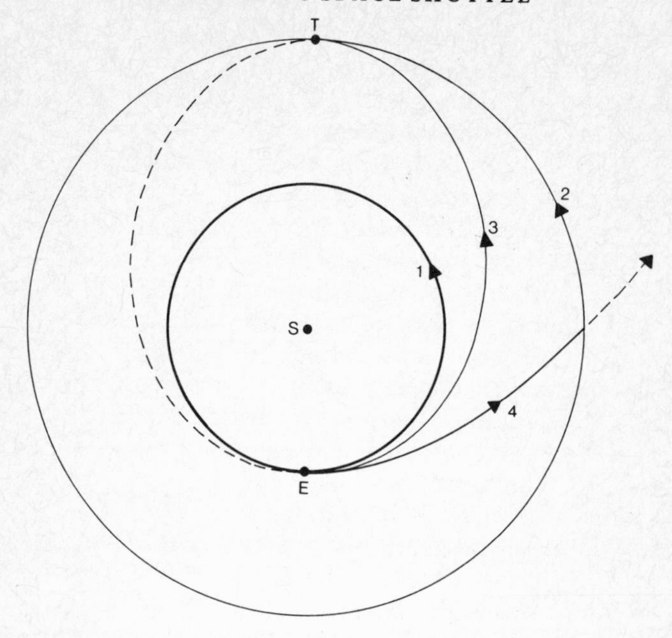

Figure 4.

Transfer orbits. To travel from the orbit of the Earth (1) to the orbit of a target planet (2) with the least expenditure of energy, the spacecraft must follow a Hohmann transfer orbit (3) – an ellipse relative to the Sun (S) which just touches the Earth's orbit at point E and the orbit of the target planet at T. To accomplish the mission in a shorter period of time a faster trajectory can be chosen (4), but to to place a spacecraft on such a flightpath requires a greater amount of energy.

orbit, the spacecraft must travel half-way round the ellipse; a typical flight to Mars would take nine months or so.

The time at which the launching takes place must be carefully chosen to ensure that the spacecraft reaches the required point on the orbit of the target planet at the same time as the planet itself gets there (see Fig. 5). In the case of Mars, the launching must take place when Mars is still well ahead of the Earth, but the spacecraft will not arrive at Mars until well after opposition (the point of closest approach between Mars and the Earth) by which time the Earth will have overtaken Mars and pulled well ahead. Opportunities for a successful Mars launch are restricted to periods of at most a few weeks which recur at intervals of, on average, 778 days. The time interval between two successive similar alignments of the Earth

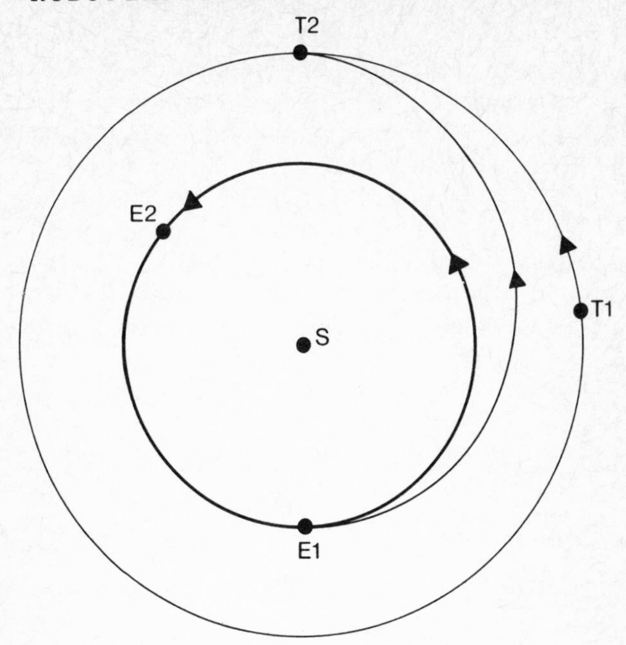

Figure 5.

Launch window. A spacecraft launched from the Earth to an outer planet along a Hohmann orbit must be launched when the Earth is at E1, some distance behind the slower-moving target (T1). If the mission planners have got their sums right the spacecraft and its target will meet at T2, by which time the Earth will have overtaken the target and moved ahead to point E2. Suitable alignments occur periodically (see text).

and a planet is called the 'synodic period', and this period determines the interval between usable 'launch windows'.

Mission planners are able to play a remarkable game of 'interplanetary billiards' in which they 'bounce' the spacecraft from the gravitational field of one planet, on towards another. By making use of each encounter to add or subtract velocity to or from the spacecraft, it can be made to fly further out, or drop closer in, as required, without additional expenditure of fuel. Just how this gravitational 'slingshot' effect can be made to work to great advantage is described below.

If a spacecraft were despatched from Earth with just sufficient velocity to reach the orbit of Jupiter, or a little way beyond, by the time the craft reached Jupiter's vicinity, it would have lost most of its initial velocity and would be moving relative to the Sun more slowly than the giant planet. As Jupiter was catching up with the spacecraft, the craft would fall under Jupiter's powerful gravitational influence, accelerate towards the planet and pass by on an hyperbolic path, eventually to recede once more and escape.

If Jupiter were stationary, the extra speed gained during the approach would be lost again as the spacecraft receded. But, because Jupiter is travelling in its orbit at a speed of about 13 km/sec. the spacecraft ends up moving off in much the same direction as Jupiter itself is travelling. Its velocity relative to the Sun – which is what matters – will be roughly equal to the relative velocity between the spacecraft and Jupiter when first it began its approach to the planet, *plus the velocity of Jupiter in its orbit*. The precise gain in velocity attained by the spacecraft depends upon its initial velocity, the angle of approach and the distance at which the craft made its closest approach to Jupiter, but the gain can amount to a large fraction of Jupiter's 13 km/sec.

TABLE 2

Launch velocities and final velocities

Initial velocity from the surface of the Earth (km/sec)	*Final velocity after 'escape'* (km/sec)
11	0
12	4.8
13	6.9
14	8.7
15	10.2
16	11.6
17	13.0

N.B. To escape from the Solar System a spacecraft launched in the direction of the Earth's motion requires a velocity 'after escape' from Earth of just over 12 km/sec.

After the encounter, the spacecraft enters a new orbit relative to the Sun, which may be an ellipse taking it much further out from the Sun, or even a hyperbola, which takes the spacecraft out of the Solar System altogether. Encounters such as this may be used to increase or to decrease spacecraft velocities, to change their directions of motion and propel them on to other targets without further expenditure of fuel (except for minor course corrections). A close encounter with Jupiter in December 1973 accelerated the American spacecraft Pioneer 10 to beyond the escape velocity of the Solar System; this spacecraft thus became the first interstellar probe, and carries a plaque showing where it came from just in case 'someone out there' should happen to find it! The interplanetary billiards technique was used on two occasions in 1974 – to send Pioneer 11 on towards Saturn after a Jupiter fly-by, and to send another American spacecraft (Mariner 10) to Mercury by way of a Venus encounter. Since that time the technique has been used to good effect on a number of occasions.

Returning the data

Data obtained by the planetary spacecraft is returned to the Earth by radio, the feeble signals being picked up by large radio antennæ such as the three 64-metre 'dishes' of NASA's 'Deep Space Network'. Pictures are transmitted in the following way. Just as a newspaper photograph is made up of a large number of little dots of different shades of grey, so each picture taken by the onboard cameras is divided up into a number of little boxes, or 'pixels' (picture elements). The degree of darkness of each pixel is represented by a binary number (the binary system counts on a base of two rather than the base of ten used in the decimal system; binary numbers are represented by combinations of the two digits 0 and 1) and the data is transmitted in binary code, where each binary digit represents one 'bit' of information.

For example, on the Mariner 4 Mars mission of 1965, each picture comprised 40,000 pixels, and each pixel was described by a six-digit binary number representing 128 possible shades of grey, so that the total number of bits per picture was 240,000. The pictures, once taken, were stored on board and then transmitted at the leisurely 'bit rate' of 8.3 bits per second. It took 8.3 *hours* to transmit each picture. As a measure of the rate of progress in telemetry (the science of transmitting information over a distance) it need only be said that the Voyager spacecraft which passed Jupiter in 1979 (fourteen years after Mariner 4's mission) returned data at a bit rate 14,000 times faster – 115,200 bits per second – and from a much greater distance, too. Each Jupiter picture contained 5 million bits, yet was transmitted in less than 45 seconds at a maximum transmitter power of 28 watts – less power than a bedside lamp!

The Goddess of Love has a hostile disposition

Venus was the first planet to be examined at close range by man-made spacecraft. Prior to the Space Age, our knowledge of that planet was decidedly limited, despite the fact that Venus approaches closer to the Earth than any other planet. There were two main reasons for this: firstly, because Venus is closer to the Sun than we are, it goes through a cycle of phases, rather like the Moon, depending on the angle between Sun, Venus, and Earth. When Venus is at its closest – and it can approach to within 40 million kilometres – it lies practically between the Sun and the Earth and its unilluminated hemisphere is turned towards us. Second, Venus is completely enveloped in cloud; no surface features can be seen, and through Earth-based telescopes the best that can be seen is a few faint, ill-defined shadings in the clouds.

It was known that Venus is similar in size to the Earth and, since it lies at about 72 per cent of the Earth's distance from the Sun, it seemed reasonable to suppose it might be a little hotter. However, we had no idea of what the surface was like, nor what the temperature was (although estimates had been made), nor much of a clue as to the atmospheric composition; there was plenty of reasoned speculation, however! Even the rotation period and tilt of the axis were unknown. Estimates of the rotation period ranged between about 22 hours and over 225 days (225 days is the length of the Venusian 'year'); spectroscopic measurements made in 1956 had hinted at the possibility that Venus might rotate on its axis in the opposite direction ('retrograde') to the other planets. The situation was very confused.

Speculation was rife as to what Venus was like below its all-enveloping cloud layer. Was Venus a dry, dusty world, or was it – as had been suggested in 1954 by F. L. Whipple and D. H. Menzel – covered with vast oceans of water? Spectroscopic observations showed the presence of the heavy gas carbon dioxide, but no-one knew the proportion of this gas relative to other possible constituents. Were the clouds made of water droplets, or were they perhaps droplets of oil as Fred Hoyle had suggested in 1955? In the early part of the twentieth century there was a strong body of opinion which suggested Venus might be the abode of life, that its surface might support thick luxuriant vegetation in a steamy humid environment not unlike the tropical forests of the carboniferous period of 200 million years ago here on Earth. Although these glamorous ideas had gone out of fashion by the nineteen-fifties, the idea of a water-covered Venus, perhaps supporting primitive organisms, was not considered at all absurd.

The spacecraft changed all that. Mariner 2 provided the first shocks in 1962 when it indicated a surface temperature of about 425°C (about 700 K), far higher than had previously been believed and which, if correct, clearly ruled out the possibility of a water-covered Venus. It showed that Venus, unlike the Earth, has practically no magnetic field, and confirmed the suggestion that Venus has a slow retrograde rotation.

During 1962 the Soviet Union launched several unnamed payloads which were believed to be Venus probes but which failed to leave Earth orbit. Zond 1, launched on 2 April, 1964 met with a similar fate to that of Venera 1, communications being lost six weeks into the mission. More Soviet attempts followed.

Venera 2, at 963 kilograms about 50 per cent heavier than Venera 1, bypassed Venus at a range of 24,000 kilometres on 27 February, 1966, but again a communications failure prevented the return of any data. Two days later, however, Venera 3 scored a bull's eye, striking the surface near the centre of the disk and so becoming the first spacecraft to reach the surface of another planet. Once again, a last-minute communications failure prevented the return of any useful data, and the Venus programme was turning out to be for the

Russians what the Ranger programme had been for the American Moon effort.

Persistence paid off when Venera 4 succeeded in parachuting a 383-kilogram package into the Venusian atmosphere on 18 October, 1967, transmitting information back to Earth as it descended. The following day the American spacecraft, Mariner 5, flew by at a range of 3,946 kilometres sending back a variety of data, including temperature and atmospheric composition (the data indicated 72-87 per cent carbon dioxide). At first it had been thought that Venera 4 had continued transmissions until it had landed; its results indicated a temperature of about 280°C and an atmospheric pressure of some 18 Earth atmospheres at 'ground' level. Comparison with Mariner data gave a clear discrepancy, and it was eventually concluded that Venera had ceased to transmit at an altitude of about 26 kilometres. Extrapolation of the results led to much higher ground temperatures and pressures in line with the Mariner results.

Veneras 5 and 6 entered the planet's atmosphere in May 1969, transmitting data during their descent, but ceasing to transmit before reaching the surface. The message was clear – conditions on Venus are so hostile that not even robot spacecraft can endure them. Nevertheless the Venera 6 results showed that the carbon dioxide content of the atmosphere amounted to between 93 and 97 per cent and confirmed that the surface pressure must be about 100 Earth atmospheres.

A successful soft landing was finally achieved on 15 December, 1970 when Venera 7, after suffering violent deceleration (of about 400 g) and severe heating (to temperatures comparable to the surface of the Sun) during its plunge through the atmosphere, touched down on the Venusian surface. It continued to transmit for 23 minutes before succumbing to the awful conditions. Venera 8 surpassed this performance in 1972, continuing to transmit from the surface for 107 minutes, and supplying data on, for example, the radioactive mineral content of the surface rocks; these results indicated that the rocks were similar to terrestrial granite.

Mariner 5 flew by on 5 February, 1974, while en route to Mercury, and transmitted the first close-range pictures of the

Venusian clouds. Ultraviolet pictures showed clearly that the atmosphere rotates round the planet in a period of about 4 days – a result which had been obtained by Charles Boyer in France back in 1957, although not widely accepted at that time.

The Soviet Venera series continued. Veneras 9 and 10 landed capsules on the surface in October 1975, each of which transmitted one photograph showing the rock-strewn terrain and demonstrating that, despite the dense cloudy atmosphere, enough light filters down to the surface to provide a level of light which has been described as being comparable to that in Moscow at noon on a dull winter's day (the temperature, however, would be strikingly different!). Veneras 9 and 10 also placed orbiters around the planet to measure surface elevations by radar.

Towards the end of 1978 a small armada of spacecraft approached Venus, the Russian Veneras 11 and 12 and the American craft Pioneer Venus 1 and 2. Pioneer Venus 1 went into orbit, carrying out radar mapping, while the latter released five atmospheric probes, one of which transmitted from the surface for 67 minutes. Most recent of the Venus missions have been Veneras 13 and 14, which landed instrumental modules on the planet in March 1982 and returned colour photographs of a rocky terrain under an orange sky.

As a result of the spacecraft explorations we now have a radically different and far more comprehensive view of the planet than we possessed at the beginning of the Space Age. Venus is unutterably hostile, its rocky surface so hot, at 480°C (about 750 K), that it glows, its stifling atmosphere (97 per cent of which is carbon dioxide) exerting a pressure of 91 Earth atmospheres at ground level, and its main cloud layer (between altitudes of 48 and 57 kilometres) made up of droplets of sulphuric acid. A visitor to Venus would be crushed, dissolved and incinerated! Radar images reveal the presence of a vast rolling plain and of 'continents'. There are two major volcanic regions, apparently active, above which have been detected flashes of lightning. Venus's crust seems to be thicker than that of the Earth, and continental drift seems unlikely.

Venus's dense carbon dioxide atmosphere acts as a greenhouse to trap heat and maintains the temperature far higher than it would otherwise be. It may well be that Venus had oceans of water in the past, but water evaporated to the atmosphere enhanced the ability of the atmosphere to retain heat and pushed the temperature up still further. As a result, the oceans were evaporated completely by a 'runaway greenhouse effect' and largely lost into space, leaving behind the torrid barren world we see today. As, with industrial activity, we modify the Earth's atmosphere today, Venus serves to remind us of what the Earth could become if we upset the ecological balance too severely.

To the Red Planet

Mars, named after the Roman god of War, is strikingly obvious in the night sky because of its distinctive red colour. Although only half the size of the Earth, it is a fascinating world and, seen through a telescope, much more rewarding than Venus. Its atmosphere is transparent (apart from occasional dust storms) and permanent dark markings have been seen on its surface since the earliest days of telescopic observation. White caps at north and south poles, which changed with the seasons, were visible and it was speculated that these were sheets of water ice like the terrestrial polar caps.

For a long time Mars was considered the most likely planet – apart from the Earth itself – to support life. In 1877 the Italian astronomer Giovanni Schiaparelli reported seeing a number of thin linear features on its surface, and these he termed 'canali', or 'channels' a term which came to be translated into English as 'canals'. Some observers claimed to be able to see complex networks of these features – others could see none. The great American observer Percival Lowell was convinced that the canals were artificial, the product of a race of advanced beings which had built the canal system to transfer water to the arid equatorial regions. Books like H. G. Wells' *The War of The Worlds* had done much to fix in the public imagination the idea that Mars was the abode of life.

As the twentieth century progressed, astronomical measurements made it seem less and less likely that Martians would exist, but the existence of vegetation was still considered plausible in the early nineteen-sixties. Prior to the spacecraft era the atmosphere was considered to be composed mainly of nitrogen (like our own) and to have a ground pressure of about 85 millibars (between 8 and 9 per cent of the terrestrial value), comparable with our atmosphere at an altitude of 17-18 kilometres.

With the tantalizing question of the existence or otherwise of life, it is not surprising that Mars was high on the list of priorities for spacecraft investigations. Once again the Soviet Union made the first attempts. First of these to get away from the Earth was Mars 1, a 900-kilogram spacecraft launched on 1 November, 1962. Five months later, at a range of 106 million kilometres, communication was lost, but it is believed that Mars 1 missed its target by about 190,000 kilometres.

The November 1964 launch window was a busy one which saw the launching of the American spacecraft Mariners 3 and 4, and the Russian craft Zond 2. Communications were lost with Zond 2 five months into the nine-month mission, which is a great pity, as the craft is believed to have passed by Mars at a range of 1500 kilometres in August 1965. Mariner 3 failed to go anywhere near Mars, but Mariner 4 met with success. After a flight lasting 228 days, Mariner 4 flew past its target on 14 July, at a range of 9,850 kilometres, sending back 21 photographs which showed clearly that Mars has its share of craters.

A most important achievement was the radio occultation experiment: as Mariner passed behind Mars, its radio signal passed through the martian atmosphere, and measurements of the attenuation of this signal showed that the atmospheric pressure at ground level was only about one tenth of what had previously been assumed. The martian atmosphere was shown to be very tenuous and clearly could offer little or no protection against harmful ultraviolet radiation: the prospects for life on Mars were receding fast.

When the pictures came back via the giant 64-metre radio dish at Goldstone, California, there was a sense of anti-climax among some astronomers. To get the pictures back was an outstanding technical achievement, to find craters on Mars was of great scientific interest, but there was a feeling of 'Not another cratered world!' which led some to question the wisdom of sending further probes in that direction simply to investigate another Moon-like world. With the benefit of hindsight it turns out that Mariner 4 flew over one of the least interesting parts of the martian surface. Fortunately, other missions did go ahead and revealed that Mars has a fascinating surface topography, like the Moon in some respects, like the Earth in others, and unlike either of them in some of its facets.

The Mariner 4 photographs revealed details down to about 3 kilometres in diameter and in the best photographs, features smaller than one kilometre across were seen. Mariner 4 achieved a similar improvement in resolution compared to Earth-based views of Mars, to that which Galileo's telescope attained for the Moon when compared to the naked-eye view.

Mariners 6 and 7 flew by in 1969, returning between them over 200 pictures, each containing about a hundred times as much information as a Mariner 4 picture. The next launch window, in May 1971, provided the opportunity for launching Mariners 8 and 9, and the two Soviet craft, Mars 2 and Mars 3. They suffered mixed fortunes. Mariner 8 dropped back ignominiously into the Atlantic Ocean. Both of the Russian craft reached Mars, on 27 November and 2 December, both entered orbit, and both released capsules to land on the planet. The Mars 2 lander crashed, while the Mars 3 lander transmitted from the surface for only 20 seconds before contact was lost. Speculation was rife that a martian had come along and switched it off!

Mariner 9, although last of the four to be launched, followed a faster trajectory, arriving at Mars on 13 November and entering an elliptical orbit round the planet to become its first man-made satellite. There was consternation when the cameras were first switched on, for Mars appeared as a blank, almost featureless disk on which nothing of significance could be seen except four dark spots. The arrival of Mariner 9 had coincided with one of the major planet-wide dust storms which are occasionally whipped up by martian winds.

The cameras were turned on the small rocky satellites. Phobos and Deimos, which turned out to be irregular bodies, pitted with craters, approximately 25 kilometres and 15 kilometres respectively in diameter.

After a few weeks the dust began to settle and the nature of the four dark spots became apparent – they were the tops of four giant volcanoes, the largest structures of this kind yet seen in the Solar System. The largest of them was named Olympus Mons (Mount Olympus – in honour of the abode of the ancient Greek gods) and it proved to be similar to a terrestrial 'shield' volcano, but on a much grander scale than anything seen on Earth. Olympus Mons reaches up to 25 kilometres above the mean surface level, has a summit crater 65 kilometres across, and covers an area some 600 kilometers in diameter. By contrast, the largest terrestrial equivalent, Mauna Kea in Hawaii stands only 9 kilometres above the ocean floor. Whether or not the martian volcanoes are completely extinct is not known, but it is likely that they have played a major role in martian history.

The Mariner 9 pictures revealed a world of fascinating variety. Apart from the craters and the volcanoes, there were depressed basins, akin to the lunar mare, and valleys of various shapes and sizes, including the gigantic chasm, Vallis Marineris, which runs for nearly 4,000 kilometres with a maximum width of about 200 kilometres and a greatest depth of nearly 7 kilometres. By comparison, the Grand Canyon in Colorado appears puny indeed.

There are 'fluvial' features, some of which look like dried-up river beds, and others – 'teardrop features' – which look like material deposited round obstructions by the action of running water. The evidence suggests that although there is no running water on Mars now, there may have been significant amounts in the past; perhaps the water came from the volcanoes.

The Soviet Union made a concerted attack on Mars at the next launch window in 1973; four spacecraft, Mars 4 to 7 inclusive, were launched, but none of them achieved the hoped-for successful landing. At the next opportunity, in 1975, the Americans launched two 'Viking' spacecraft with the aid of powerful Titan III–Centaur launch vehicles. Each spacecraft consisted of an orbiter, weighing some 2300 kilograms, and a lander – a three-legged structure with a mass of some 600 kilograms. Viking 1 entered orbit round Mars on 19 June, 1976 and was joined on 7 August by Viking 2.

After the orbiter had carefully surveyed potential landing sites, the Viking 1 lander separated and fired its retro-rocket to enter the martian atmosphere and attempt a landing. An aerodynamically shaped shield protected the lander from atmospheric heating as the vehicle plunged into the atmosphere at a speed of about 16,000 km/hour then, having finished its task, it was ejected. At an altitude of about 6 kilometres a parachute opened and about a minute later, at a height of just over 1 kilometre, three descent engines fired to slow the craft to a gentle impact at a speed of about 9 km/hour.

On 20 July, 1976 – exactly seven years after the Apollo 11 lunar landing – Viking 1 touched down in a relatively flat but rock-strewn region known as Chryse Planitia, at martian latitude 22.5° N. and longitude 48° W. Immediately it set to work with its cameras, meteorological instruments, and surface sampler. Some six weeks later, on 3 September, Viking 2 touched down at latitude 48° N, longitude 226°. The Viking mission had scored two out of two.

While the orbiters continued the detailed mapping and investigated the global properties of the atmosphere, the landers carried out experiments to investigate meteorology, photograph the surface, and to investigate the physical, magnetic, chemical, and biological properties of the surface layer. One of the prime aims of the mission was to settle the burning question of whether or not life exists on Mars. Initially two of the three relevant experiments gave results which could be interpreted as indicating biological activity in the martian soil, but the third revealed no evidence of any organic material remotely resembling even the most basic living material here on Earth. As confusion mounted over the interpretation of the continuing measurements, the conclusion reluctantly arrived at was that the results of the earlier experiments must have been due to some exotic martian chemistry rather than to genuine biological activity.

There the matter rests at present. Life on Mars is not completely ruled out, but there are no large organisms or plants, and certainly no intelligent martians!

The atmosphere is tenuous, but the polar caps – as the Viking orbiters showed – are made up principally of water ice. Just how much water is there is uncertain, but probably no

Plate 6. The surface sampler of the Viking 1 spacecraft seen against the rock-strewn surface of the Chryse Planitia on Mars. (NASA)

more than enough to cover the whole surface to a depth of a few metres if the caps were melted. Certainly there is nothing like the content of the Earth's oceans.

Mars remains a fascinating world, chilly and barren, but with a fascinating geology. Although less hospitable than had been hoped a few decades ago, it is still the least hostile of the planets in the Solar System, and merits further investigation.

Mercury

The other terrestrial planet, Mercury, has been visited by only one spacecraft so far – Mariner 10. Using a close encounter with Venus to swing it closer to the Sun, Mariner 10 made its first close fly-by of Mercury at a range of 5,790 kilometres on 29 March, 1974. Careful juggling of its trajectory allowed the spacecraft to loop round the Sun and return to the vicinity of Mercury on two more occasions before its attitude control gas ran out and the instruments were switched off.

Mariner 10 returned nearly 2000 photographs and showed Mercury to be a heavily cratered world, like the Moon. That this tiny, airless world should be heavily cratered came as no great surprise, but the discovery of a magnetic field of about 3 per cent of the strength of the Earth's field caused a fair degree of astonishment. Despite its small size, Mercury's magnetic field proved to be stronger than that of Venus or Mars. Mariner also picked up evidence of an exceedingly tenuous 'atmosphere' – not a permanently attached envelope of gas, but made up of hydrogen and helium flowing from the Sun which becomes temporarily bound to the planet.

Hostile and barren, and suffering extremes of temperature, this intriguing little world nevertheless has a fascination of its own.

Through the asteroid belt

Before Mariner 10 had begun its voyage to Mercury, NASA turned its attention to the giant planet Jupiter, a world utterly different from the Earth, which presents a fascinating and ever-changing panorama when seen through Earth-based telescopes which readily reveal the turbulent cloud belts at the top of its deep atmosphere. Despite its great bulk Jupiter spins rapidly on its axis, its rotation period being 9 hours 50 minutes. As a result of its rapid motion it bulges out appreciably at the equator and is flattened at the poles. The rapid rotation has dragged out its weather systems into an alternating pattern of dark belts and lighter zones of clouds, swirling and ever-changing in appearance, the only semi-permanent feature being the Great Red Spot, a reddish oval covering a region up to 30,000 kilometres long and 10,000 kilometres wide – considerably larger than the surface area of the Earth.

As the largest planet in the Solar System it has a retinue of satellites including four large moons first seen by Galileo when in the winter of 1609–10 he first turned his primitive telescope to the skies and initiated a revolution in observational astronomy. A new revolution in understanding the giant planets began with the launching of the American spacecraft, Pioneer 10, in 1972.

The 270-kilogram spacecraft carried eleven different instruments to carry out thirteen major experiments. A highly sophisticated craft it was capable of returning information from Jupiter at a rate of 1024 bits per second via its 2.7-metre dish-shaped antenna with a transmitting power of only 8 watts. Its transmissions were received by the three 64-metric radio dishes of the Deep Space Network in California, Spain, and Australia; these in turn were able to beam commands to the spacecraft with powers of up to 400,000 watts, so as to make the commands detectable at a range where radio waves take an hour or more to travel from Earth to spacecraft.

Previous interplanetary probes had used solar panels to produce electrical power, but where Pioneer was headed solar radiation would be too weak for this purpose. Instead Pioneer 10 became the first planetary explorer to rely wholly on nuclear-electric power, heat from the radioactive decay of Plutonium-238 acting on banks of thermocouples generated enough power to keep the spacecraft operating in the distant depths of the Solar System.

Pioneer 10 blasted off on 3 March, 1972, surging up to a velocity of 51,700 km/hour – unprecedented at that time – and passing the Moon *en route* after only 11 hours. Within three months it had crossed the orbit of Mars and was heading into unknown territory. The big question at that time was, 'Would Pioneer 10 get through the asteroid belt unscathed?' No-one knew for sure how many asteroids lay between Mars and Jupiter, nor if there were myriads of much smaller particles which could strike the spacecraft a fatal blow. Even particles the size of a grain of sand could do significant damage.

Its progress was anxiously monitored as it entered the main body of the asteroid belt in July and emerged again, completely unscathed in February, 1973. A major hurdle to planetary exploration had successfully been cleared. On 3 December, after a voyage lasting 21 months and spanning a billion kilometres, Pioneer skimmed past Jupiter's cloud tops at a range of 130,345 kilometres and a speed of 132,000 km/hour. Apart from taking beautiful photographs, Pioneer's instruments scanned the planet and its environs and, as it plunged deep into Jupiter's magnetosphere, the level of radiation in Jupiter's equivalent of the Van Allen zones, climbed alarmingly, saturating the instruments. No man could have survived the hostile radiation environment through which Pioneer struggled.

As a result of its encounter Pioneer 10 picked up sufficient velocity to ensure that it will escape altogether from the Solar System.

Its identical twin, Pioneer 11, made an even closer approach to Jupiter on 3 December, 1974, reaching a maximum speed of 160,000 km/hour as it skimmed within 43,000 kilometres of the cloud tops. Its trajectory took the spacecraft under Jupiter's south pole and back over the north pole, so minimizing the time spent in the intense radiation zone. This flightpath sent Pioneer 11 out of the ecliptic (the plane of the planetary orbits), across the orbit of Jupiter and beyond to a rendezvous with Saturn on 1 September, 1979. After a three-billion-kilometre journey lasting six years Pioneer 11 – still functioning normally – passed by the ringed planet at a range of 21,000 kilometres and returned photographs (rather disappointing in their lack of detail) and a host of other information. Pioneer 11 discovered two more rings, showed that Saturn has a powerful magnetic field, and revealed that – like Jupiter – Saturn emits considerably more heat than it receives from the Sun. Clearly both of these planets must have very hot interiors.

Most ambitious of all the deep-space probes so far have been the two Voyagers. More advanced versions of the Mariner class of spacecraft, each Voyager weighed 825 kilograms and carried the largest size of communications antenna yet to fly on a planetary probe – 3.7 metres in diameter. Pointed towards the Earth the antenna enabled the spacecraft to transmit data at a rate 25 times faster than the Pioneers had attained (115,000 bits per second at Jupiter and 44,800 bps at Saturn). Three radio-isotope thermoelectric generators provided electrical power, and these were mounted at the end of a boom to keep them as far as possible from the 11 scientific instruments.

Voyager 2 was launched on 20 August, 1977, sixteen days before Voyager 1 which was scheduled to follow a faster trajectory and arrive at Jupiter four months ahead of its twin. On 14 October it looked as if disaster had struck. Communication was lost when the star sensor – which was supposed to lock on to the bright star Canopus in order to keep the antenna pointing towards the Earth – locked on to the wrong star. To the relief of all concerned, communication was re-established two days later. Although similar problems arose again later in the flight, all was set fair for the Jupiter encounter when Voyager 1 began its seven-month observational programme on 6 January, 1979. Three months later, on 5 March, Voyager 1 made its closest approach. Voyager 2 followed suit on 9 July.

Jupiter's gravitational slingshot hurled both Voyagers onward towards their next port of call – Saturn.

The results obtained at Jupiter by the Voyagers were spectacular in the extreme and would require a complete book to do them justice. High resolution colour photographs revealed the majestic beauty of the planet for all to see, and there were surprises galore. Jupiter was shown to have a ring – much more modest than the magnificent Saturnian ring, admittedly – 128,000 kilometres in radius but only 30 kilometres wide. The complex circulation of the clouds and atmospheric currents was revealed in detail, and the Great Red Spot shown to be elevated above the mean cloud level; it seems to be a high-pressure area round which strong circulation occurs, its red colour possibly resulting from the presence of phosphorus. Lightning bolts and auroral arcs were seen in the Jovian atmosphere.

The greatest surprises of all came with the close-up view of the four major satellites, in order of distance from the planet, Io, Europa, Ganymede, and Callisto. Each is a globe comparable with or larger than our own Moon, and each is utterly different from the others. Callisto proved to be the most cratered world seen so far, every point on its icy surface saturated with impacts. Ganymede showed two distinct types of terrain in opposite hemispheres, a heavily cratered region and a smoother grooved region indicative of melting processes which had obliterated the craters and tensional forces which had produced the grooves. Europa seems to be the smoothest body in the Solar System, with practically no vertical relief, the surface being covered with shallow intersecting cracks.

Strangest of all was Io, which turned out to be covered in

Plate 7. Jupiter's Great Red Spot photographed on 1 March, 1979 from a range of 5 million kilometres, by the Voyager 1 spacecraft. (NASA)

violently active volcanoes hurling plumes of material to heights in excess of 250 kilometres. This was a complete and utter surprise, for practically no-one had expected to find active vulcanism on a satellite. The current theory to explain this behaviour is heating by tidal pumping. As Io moves round Jupiter, its orbit is periodically perturbed by the other moons, in particular Europa and Ganymede. The result of these conflicting forces is to squeeze and stretch Io, raising tidal bulges in the body of the satellite perhaps 100 metres high. The energy released in this way keeps Io's interior in a molten state and sustains volcanic activity at such a pitch that periodically, Io is virtually turned inside out as a new surface builds up and the old surface sinks down into the interior. Its sulphurous surface revealed in photographs gives it the appearance of a diseased pizza!

Jupiter's magnetosphere – the region of space within which its magnetic field dominates that of interplanetary space, and within which atomic particles are trapped – is about a hundred times larger than that of the Earth. The effect of the solar wind is to compress the magnetosphere on the sunward side and stretch it out into a long tail – the magnetotail – on the far side. On the sunward side Jupiter's magnetosphere extends typically to some 5 million kilometres, but the magnetotail stretches beyond the orbit of Saturn – a distance of over 600 million kilometres. Indeed Jupiter's magnetotail has been likened in many respects to the tail of a comet.

With unerring accuracy the two Voyagers marched onwards to their appointed meetings with Saturn, Voyager 1 making its closest approach at a range of 124,000 kilometres on 12 November, 1980. The Voyager 1 trajectory was planned to allow a close fly-by of the giant satellite Titan, a world so interesting in its own right, that, if Voyager 1 had failed to return adequate data, Voyager 2 would have been retargeted to make a Titan encounter and would then have been unable to use Saturn's gravity to place it *en route* for Uranus – the next stage in the Grand Tour of the planets. Voyager 1 performed flawlessly and showed once again – although space scientists were by now becoming used to this kind of thing – that preconceived ideas based on telescopic observations were wrong in a number of respects, particularly with regard to Titan's atmosphere. Titan was known to have an atmosphere (and was the only satellite to have one); this atmosphere was thought to be comparable in density with our own and to be composed primarily of methane. It was thought that Titan's atmosphere might resemble the primitive terrestrial atmosphere at the time life was beginning to emerge on Earth. Nothing could have been further from the truth – Titan's atmosphere was shown to consist mostly of *nitrogen*, although, admittedly, there is some methane present. The mean temperature of the surface is about −180°C (about 90 K) and in the polar regions there are probably seas of liquid nitrogen. The surface is shrouded in a dense haze at least 280 kilometres thick. Once considered to be the largest of the planetary satellites, Titan, with a diameter of 5,100 kilometres, proved to be a little smaller than the largest Jovian satellite, Ganymede.

Plate 8. The grooved rings of Saturn revealed by the Voyager 1 spacecraft in November, 1980. (NASA)

As for Saturn itself, the Voyagers revealed a magnetosphere about one-third the size of the Jovian one, and showed that the disk is rather more bland in appearance than that of Jupiter, although belts and spots do exist in profusion. Wind speeds over 1600 km/hour were noted, four or five times stronger than those observed on Jupiter. The satellites, too, showed a variety of interesting features, and Saturn's tally of Moons was shown to be possibly as high as 23.

But, of course, it was the ring system which provided the greatest interest and the most surprises. Hundreds of rings and divisions were revealed; the system had more grooves than a long-playing record. Even the so-called Cassini division – seen from Earth as a dark gap – turned out to contain several narrow ringlets. A faint innermost ring – the D-ring –

was revealed to extend almost down to the cloud tops, and the strange F-ring lying just outside the main system, and have a radius of 140,000 kilometres was shown to consist of at least three strands, two of which were intertwined – braided like a rope. Two tiny moons seem to act like 'shepherds' to keep the F-ring particles marshalled in position.

Dark spoke-like structures in the brightest B-ring seemed to defy the normal laws of motion and are now believed to be due to fine, electrically charged, dust particles, suspended above the main body of the ring system. All in all, the Saturnian system has proved to be one of fascinating complexity which will keep astronomers analysing, interpreting, and theorizing for decades to come.

The Voyager missions mark the pinnacle of the planetary exploration programme in the first 25 years of the Space Age. All five of the planets which can be seen with the naked eye – and which have fascinated mankind since the dawn of history – have been investigated *in situ* by man-made spacecraft. Between the first Venus encounter and the second Voyager Saturn encounter, less than twenty years have elapsed, and in that period of time we have learned perhaps a hundred times as much about the planets as had been discovered in the whole of recorded history. Some ideas have been confirmed, many more have been overthrown, and a host of phenomena, previously undreamed of have been revealed by these willing robots.

The odyssey has found its 'Golden Fleece' many times over. What new wonders lie in store?

6

New Windows on the Universe

To the astronomer, the Earth's atmosphere is a nuisance. It is vitally important for our survival, of course, but it is a great impediment to our study of the universe because it is opaque to most forms of radiation. Light is a form of electromagnetic radiation, an electric and magnetic disturbance which travels through space at the velocity of light – 300,000 kilometres per second – and which behaves like a wave motion. Thinking of light as being like a wave on water, we can think of its wavelength as being the distance between two successive wavecrests; we can also talk about the *frequency* of the wave as being the number of wavecrests per second which pass an observer. Light is a curious phenomenon. Although it behaves like a wave in some ways, in other respects its behaviour is better described by thinking of light as being a stream of tiny particles, called photons. In practice we choose whichever description is more convenient for the particular investigation in hand.

The wavelength of visible light is less than one millionth of a metre, and is often described by means of a unit known as the nanometre (one nanometre [nm] is a thousand millionth of a metre). Our eyes respond to different wavelengths of light by recognizing different colours; thus red light has a longer wavelength (about 700 nm) than blue light (about 400 nm), and other colours correspond to intermediate wavelengths. Light of these wavelengths can penetrate the atmosphere to ground level where it can be collected by telescopes and studied by astronomers. However, the sky is often obscured by cloud, it is dusty and polluted, and it is turbulent, causing stars to twinkle, and the images of objects seen through telescopes to shimmer and shake as if viewed through a rippling pool of water.

To try to minimize the deleterious effects of the atmosphere major observatories and telescopes are sited on mountain tops, but only a modest improvement can be obtained in this way. The real solution to the problem is to place telescopes in orbit above the atmosphere.

Visible light is one type of electromagnetic radiation. There is a complete range of wavelengths – known as the *electromagnetic spectrum* – extending from the very short to the very long. Radiation of wavelength shorter than visible is known as ultraviolet; wavelengths shorter than a few tens of nanometres are called x-rays, while the shortest waves of all – having wavelengths of less than about 0.01 nanometres – are known as gamma rays. The shorter the wavelength, the more energetic the radiation, and the more damaging are its effects on living tissue. Wavelengths longer than visible light are termed infrared; beyond about 1 millimetre, they are termed microwaves, and beyond a few tens of centimetres are conventionally regarded as radio waves.

None of the shorter-than-visible radiation – with the exception of a very small fraction of the near-ultraviolet, reaches the ground; it is all absorbed in the upper atmosphere, atmospheric ozone being a major absorber of ultraviolet, for example. Water vapour and carbon dioxide are likewise highly effective at absorbing infrared, and only a small fraction – at certain restricted wavebands – actually penetrates to ground level or even to mountain-top observatories.

In the microwave and radio part of the spectrum, there is a range of wavelengths from about a centimetre to between 10 and 20 metres which can penetrate the atmosphere, but longer waves are absorbed or reflected back to space. The visible range of wavelengths which get through the atmosphere constitute the optical 'window', the radio wavelengths, the radio window. From the dawn of time Man has observed the universe through the optical window with the aid of his eyes alone. The development of the optical telescope since the early seventeenth century has immeasurably enhanced his ability to gaze through that particular atmospheric window. In 1933 the American radio engineer Karl Jansky discovered that radio waves are reaching us from space, but initially, little interest was expressed by conventional astronomers. During the last three or four decades, interest has blossomed, and giant radio telescopes have been constructed all over the globe to detect and analyse these radiations entering through the radio window.

Radio astronomy produced a revolution in astronomy almost as great as that which followed the invention of the optical telescope. On the one hand it provided us with different kinds of information about objects whose existence was already known, but on the other it led to the discovery of whole new species of object whose existence had not even been suspected – radio galaxies, quasars, and pulsars, to name but a few.

It came as no surprise, therefore, that when satellites were able to take instruments above the obscuring blanket of the atmosphere and gained access to the whole of the electromagnetic spectrum (and to other things besides) a rush of new data and discoveries immediately followed. The Space Age has removed the blinkers from astronomers' eyes and, even today, we are still to some extent blinking in the glare of the new 'light' available to us. New branches of astronomy have sprung up over the past 25 years, in particular ultraviolet, x-ray, and gamma-ray astronomy, Like radio astronomy, these new branches have provided additional information on known objects and have thrown up many new and puzzling sources of previously unsuspected kinds. We are living today in the midst of an astronomical revolution. In the early days of the satellite era it was only necessary to place in orbit a crude detector – of x-rays, say – to scan the sky and make a host of startling discoveries Nowadays, things are a little different. Instruments have to be more sophisticated, to analyse in more detail and discriminate more precisely; yet new discoveries continue to be made with great frequency.

In addition to studying electromagnetic radiation, satellites in orbit round the Earth, or spaceprobes in interplanetary space, have been able to detect directly cosmic rays and other particles, to measure magnetic fields, and to experience directly the bombardment by tiny particles of matter – micrometeorites – which swarm through the Solar System. They have investigated the solar wind, the stream of electrons, protons, and other atomic nuclei which flow from the Sun and blow past the planets like a wind; when these particles impinge on the atmosphere, they give rise to phenomena such as the aurora.

Satellites have also looked inward to study the Earth on a global scale. Among the first tasks of artificial satellites were the study of the upper atmosphere, the shape of the Earth, and the nature of the Earth's magnetosphere. The resisting effect of the upper atmosphere was first studied with Sputnik 1 as atmospheric drag brought it down to burn up after 92 days in orbit. Explorer 1 discovered the first of the Earth's radiation zones, and analysis of changes in the orbit of Vanguard 1 revealed that the Earth's equator is 'pear-shaped', rather than circular, a major step forward in the science of geodesy (the study of the shape of the Earth).

In one sense the near-Earth satellites were completely international, once they had been launched, for their orbital motion could be studied by anyone, and information about the extent and density of the upper atmosphere, and about the shape of the Earth, deduced by anyone who cared to analyse the observations which they had made. In the late 1950s and early 1960s a global network of amateur observers – armed with binoculars or small telescopes, and stopwatches – were able to make observations of real scientific value; it was a heady time for the amateur satellite watcher. So much data was amassed by professional and amateur observers that much of it never was, and probably never will be, analysed. This typifies one problem which space exploration has brought in its wake – the overwhelming mass of information.

To do justice to the achievements of the hundreds of scientific satellites would require many volumes, and it is only possible here to pick out a few highlights.

The Soviet Union has launched a great many scientific satellites of various types under the blanket heading of the 'Cosmos' series. Cosmos satellites are typically in the mass range 400 to 4000 kilograms and are built on a modular structure so that they can be, in effect, mass-produced, but with minor modifications as required. So far as possible standard units are used for power and communications; variations occur in the instrumentation packages. Cosmos 1 was launched on 16 March, 1961. By 17 December, 1965, the tally of Cosmos launchings reached one hundred, and the thousandth Cosmos went into orbit on 31 March, 1978. Of this total, over 50 per cent have been of military application – spy satellites, for example (see Chapter 7). The scientific satellites in the Cosmos programme have concentrated mainly on measuring the properties of the upper atmosphere, solar radiation, particles from the Sun, magnetic fields in space, and meteoroids.

The Soviet 'Elektron' series of 1964 was particularly interesting. Satellites were launched in pairs from the same launch vehicle – one member of each pair being placed in a fairly low orbit, ranging from about 400 to about 7000 kilometres in altitude, and the other into a highly elliptical orbit reaching out to more than 66,000 kilometres. The aim of the series was to study the Van Allen zones and solar radiations. Radiation studies were also the stated objectives of the 'Proton' satellites which aroused considerable interest when they were launched in 1965 because their high mass (12.2 tonnes) implied that they must have been launched by a new launch vehicle, considerably more powerful than previous Soviet launchers.

The Soviet Union was very much concerned to investigate the possible danger to cosmonauts of prolonged exposure to radiation in space.

Much attention, too, was paid to determining the rate of impacts due to micrometeorites. The most spectacular satellites to be devoted to this problem were the Pegasus series, three satellites of this type being launched in 1965 by the United States with the aid of Saturn I launch vehicle. At that time they were the heaviest payloads yet orbited by the United States. Once in orbit they unfolded wing-like sets of panels with a span of 96 metres to register the numbers and penetrating powers of these microscopic meteoroids.

The blanket term 'Explorer' was applied to a wide variety of American scientific satellites but different project names were often attached to denote their specialized applications. For example, Explorer 42 was also known as SAS-1 (first of a series of *S*mall *A*stronomy *S*atellites), and was designated 'Uhuru' – a Swahili word for 'freedom' as it had been launched from the Italian 'San Marco' platform off the Kenyan coast on Kenyan Independence Day (12 December, 1970).

A bewildering range of Explorers has been launched for a wide range of tasks, and only a few can be highlighted here. The GEOS (*G*eodetic *E*arth *O*rbiting *S*atellites), beginning with GEOS 1 (Explorer 29) launched on 6 November, 1965, greatly refined our knowledge of the gravity field of the Earth, and of the shape of the planet and its oceans. Other sophisticated satellites followed. LAGEOS (*L*aser *Geo*dynamic *S*atellite), launched on 4 May, 1976 to a near-circular orbit with an altitude of about 5,900 kilometres, was rather like a giant golf ball covered with laser reflectors. By reflecting laser beams from the satellite, the separation between observing sites could be pinned down with very high precision. The U.S. Geological Survey are using this satellite to measure continental drift and the movements of the San Andreas fault in California. LAGEOS is expected to have a useful life – before its reflectors become unusable – of about 50 years, and to remain in orbit for about 8 million years.

An ingenious series of Explorers have been used to help fill in the gap in continuous atmospheric observations which exists above the ceiling attainable by instrumented balloons (about 50 kilometres) and below the level at which satellites can remain in orbit for prolonged periods (ordinary satellites cannot survive for long in orbits with perigees much below 250 kilometres). Commencing with Atmosphere Explorer-C (Explorer 51) in December 1973, satellites of this type have carried an onboard propulsion system which allows them to swoop down to 120 kilometres and overcome drag (similar systems are used with some spy satellites).

The wider environment of Earth–Moon space has been probed by, for example, the IMP (*I*nterplanetary *M*onitoring *P*latform) series commencing with IMP–A (Explorer 18) in November 1963. These satellites pursued highly elliptical orbits (e.g. the apogee of IMP–A was at an altitude of 197,600 kilometres) to explore the magnetosphere and the effects exerted upon it by the solar wind.

The OGOs (*O*rbiting *G*eophysical *O*bservatories) and OSOs (*O*rbiting *S*olar *O*bservatories) extended our understanding of the links between the Sun and the Earth. The OSOs in particular were the first major series of satellites devoted specifically to observing the Sun itself in ultraviolet, X- and gamma-radiations. Since then (the OSO series began in March 1962) a wide range of satellites and space probes – some of them launched by NASA in cooperation with other nations – have greatly enhanced our knowledge of this, the nearest star. Particularly notable were the two NASA/West German 'Helios' spacecraft, launched in 1974 and 1976 which passed within the orbit of Mercury to probe the Sun and its environment from a minimum range of 45 million kilometres, and a cooperative venture between NASA and ESA (the European Space Agency) – the International Sun–Earth Explorers, one of which (ISEE–3) was placed in a state of uneasy equilibrium 1.5 million kilometres along the Sun–Earth line at the point where the attractions of Sun and Earth effectively balance out. Working with ISEE–1 and ISEE–2, it could anticipate the effects of fluctuations in the solar wind.

Most productive of all, so far as solar research is concerned, was the giant American manned laboratory, Skylab, which remained in orbit from 1973 to 1979 and which is described in Chapter 10.

Also of great importance has been the Solar Maximum Mission (SMM), an American satellite launched into a 574-kilometre orbit in February 1980. Designed primarily to make a concentrated attack on the nature, trigger mechanism, and effects of solar flares – violent explosive events which tend to occur in the vicinity of sunspot groups and which release x-rays and particles, some of which impinge on the terrestrial atmosphere (producing auroral displays, disrupting long-range radio communication, producing fluctuations in the Earth's magnetic field) – the spacecraft carried a wide range of optical, ultraviolet, x- and gamma-ray detectors. The idea was to correlate the spacecraft results with earth-based observations to obtain as comprehensive as possible an understanding of these most dramatic events.

The satellite also carried an instrument to measure accurately for the first time the so-called 'solar constant' – the amount of solar energy reaching the top of the Earth's atmosphere*.

Because of the changing and imprecisely known absorbing effect of the atmosphere, this quantity cannot be determined with even one per cent accuracy.

By the end of 1980, unfortunately, the 2.3-tonne satellite had lost stability and had begun to spin; it is still usable, but at poor efficiency. However, it is equipped with a grappling ring designed to be suitable for 'grabbing' by the manipulator arm on the Space Shuttle (see Chapter 11), and there are tentative plans to attempt a 'rescue' in late 1983 if the Shuttle keeps up to schedule.

A quarter century of space exploration has added greatly to our understanding of the Sun and has overturned more than a few of our previously held ideas. The Sun has been shown to be mildly variable in its output of energy, and these variations in the medium and long term may be an agency of climatic change on Earth. We have come to realize that powerful magnetic fields dominate the behaviour of matter on the surface of the Sun and in its atmosphere. The outer atmosphere of the Sun – the corona, visible directly from Earth only during a total eclipse – is concentrated into x-ray emitting regions with temperatures of up to 5,000,000 K, and is not the bland, uniform layer it was once thought to be. Indeed it has been shown to contain enormous holes out of which matter, in the form of electrons, protons, and other particles, streams away to make up the solar wind. The blustery solar wind, in turn, has been shown to squeeze up the Earth's magnetosphere on the sunward side, and draw it out into a long tail on the far side, and the intimate links between our planet and its parent star have been rendered tangible.

Satellites have also turned their 'eyes' on the distant depths of space – far beyond the Solar System with equally dramatic and startling results. So far the satellites have concentrated their efforts mainly at the short-wave end of the spectrum (ultraviolet, x- and gamma-radiation) although longer wavelengths have not been ignored. For example, Explorers 38 and

* This has a value of about 1370 watts per square metre; i.e. a surface of one square metre perpendicular to the Sun's rays – at the top of the atmosphere – will receive nearly 1.4 kilowatts of power; due to losses in the Earth's atmosphere, etc., domestic solar panels cannot do so well.

49 – the former placed in orbit round the Earth, and the latter in orbit round the Moon – carried giant sets of antennae, 230 metres long, to study terrestrial and celestial radio sources; and infrared astronomers should soon have their own satellite, IRAS, which is due to be launched just before the twenty-fifth anniversary of Sputnik 1.

The really dramatic results have come from the short-wave observations – the domain of 'high-energy' astrophysics.

The first major ultraviolet experiment was carried on a brief up-and-down rocket flight from White Sands in 1946; likewise the first measurement of a cosmic x-ray source – designated Sco X–1 because of its location in the constellation Scorpius – was made by a similar rocket flight in 1962. Sco X–1 is now known to be an x-ray emitting binary, a star with a white dwarf companion. Material from the star is dragged by gravity into a rapidly circulating disk of matter (the 'accretion disk') and becomes heated to many millions of degrees with the result that x-rays are emitted. Such is the x-ray power of this source that it emits a hundred thousand times as much energy in the form of x-rays alone as the Sun emits in the form of visible light. The nature of the source was not immediately apparent at the time, but the discovery pointed clearly to the importance of getting a satellite into orbit to scan the skies for more x-ray sources.

The American OAO (*O*rbiting *A*stronomical *O*bservatory) series was an ambitious one which began inauspiciously when contact was lost with OAO–1 two days after launching in April 1966. OAO–2, launched on 7 December, 1968 was the heaviest and most complex American scientific satellite up to that time, and among its other achievements, catalogued the brightnesses and positions of 5,000 ultraviolet-emitting stars and showed that the comets Tago–Sato–Kosaka and Bennett were surrounded by huge clouds of hydrogen about half the size of the Sun. Most successful was OAO–3, launched on 21 August, 1972 and named Copernicus in honour of the Polish astronomer Copernicus (who had placed the Sun at the centre of the universe and started an astronomical revolution) and in recognition of the 500th anniversary of his birth which was due to be celebrated the following year. OAO–3 was equipped with a 0.9-metre telescope capable of being locked on to the thickness of a human hair at a range of about a kilometre, and it concentrated primarily on ultra-violet studies. Among its many achievements it detected the presence of the element deuterium (heavy hydrogen) in interstellar clouds; the importance of this discovery is that the amount present gives crucial information on the past history and possible origin of the universe.

Ultraviolet observations have been made by various other satellites, including Dutch and European satellites, and the International Ultraviolet Explorer (IUE) launched in 1978 as a joint venture between NASA, ESA, and the British Science Research Council.

The first satellite devoted solely to x-ray observations was Explorer 42. Better known as 'Uhuru' it scanned the sky at wavelengths of between 0.6 and 0.06 nanometres and brought up to about 200 the tally of known x-ray sources. Uhuru's performance was surpassed by a British satellite, Ariel 5, which was launched by NASA on 15 October, 1974. Ariel 6 followed in 1979, but by far the most effective x- and gamma-ray satellites to date have been the NASA series of HEAOs (*H*igh *E*nergy *A*stronomy *O*bservatories), three of which were launched between 1977 and 1979. HEAO–2 launched in November 1978, and named the Einstein Observatory (in honour of Albert Einstein), was the first to carry a telescope capable of producing x-ray images of cosmic sources other than the Sun, and was the first, therefore, to show detail and structure in x-ray sources.

X-ray observations have opened our eyes to a wide range of startling and fascinating phenomena. The turbulent clouds of gas which represent the remnants of supernovæ (exploding stars) are x-ray sources and some of them contain x-ray emitting pulsars – sources of x-rays, associated with rapidly-spinning neutron stars, which flash on and off several times a second. Many of the x-ray sources have been shown to be binary systems, like Sco X–1, where material from one star is falling towards a collapsed companion (white dwarf or neutron star) and is heated to perhaps hundreds of millions of degrees.

Perhaps the most exciting of all these sources is Cygnus X–1, an x-ray emitting binary discovered by Uhuru which is widely believed to contain a black hole with an estimated mass of between 8 and 11 solar masses, towards which material from its highly luminous companion star is being dragged.

Some of the most violently active, highly luminous galaxies contain compact x-ray emitting objects in their central cores, and again it is widely believed that these sources may well be accretion disks around supermassive black holes containing hundreds or even thousands of millions of solar masses of material. Quasars, too, are powerful x-ray sources, for long considered to be the most enigmatic entities in the cosmic 'zoo'. They are very compact, much of their energy coming from a region of space less than one light-year (and even less than one light-day in some cases) in diameter, yet seem to lie at such enormous distances that they must be emitting more energy than a hundred normal galaxies. Many astronomers now believe that quasars are powered by supermassive black holes, lurking at the centres of galaxies, and digesting material from the centres of the galaxies which they inhabit; in so doing they emit so much energy that the light from the surrounding galaxy is simply overwhelmed.

X-ray emissions from intergalactic matter in clusters of galaxies have also been detected, and a general weak background of x-rays from the whole sky may be due either to widespread intergalactic matter or to a large number of individual distant sources. These observations may have a crucial bearing on the whole future of the universe. Most astronomers are of the opinion that the universe began in a hot, dense, explosive event – the Big Bang – some ten or twenty billion years ago. The universe has been expanding, and the galaxies rushing apart ever since. The big question is whether or not this expansion will continue for ever; if there is sufficient matter in the universe, the combined gravitational attraction will be sufficient eventually to halt the headlong separation of the galaxies and cause the universe ultimately to collapse in a 'Big Crunch'. If there is insufficient matter, the expansion will continue for ever, and the universe will eventually become a cold, dark, practically empty void.

At present the evidence suggests there is insufficient material to halt the expansion; even taking account of the intergalactic material found so far, there is no more than a tenth of the amount necessary to 'close the universe'.

There is no doubt that space-based observations will be crucial to resolving this great and intriguing mystery just as they have already resolved many mysteries and raised many more questions. No doubt the new eyes which satellites have provided will continue to enhance and deepen our perception of the universe in the years ahead.

7

Satellites at Work

Few people who heard the repetitive 'bleep, bleep' signal of Sputnik 1 on the day the Space Age dawned could have imagined the extent to which satellites – in a multitude of applications – would impinge on the lives of all of us. Even today, few are fully aware of the extent of the satellite activity which is going on above our heads. In the fields of communication, meteorology, earth resources, navigation, military surveillance (and potentially offensive roles) satellites are exerting a growing and, in some areas, already dominant influence. As we move into the second quarter century of the Space Age we may expect their role to expand rapidly.

Communications

Nowhere is the influence of the satellite more profound than in the field of communication. Mankind has been described as 'the compulsive communicator'; we delight in absorbing and exchanging information and opinions. But communication in its widest sense has been the key to human history, to the evolution from a nomadic past to city states, coherent nations, and larger units; perhaps eventually it may lead to a single global state. The existence of lines of communication, and the speed at which information could be communicated, were crucial to the growth and effective control of states and empires. But improving communications have also allowed the spread of ideas which have undermined the control of a state over its subjects. Like most new innovations, rapid, long-range communications have proved to be a two-edged weapon.

The horse, the wheel, the railway, the aircraft, and radio have all revolutionized the world we live in, but satellite communications have brought every part of the world within range of near-instantaneous contact and have begun a revolution which cannot be halted and which already is changing the social and political complexion of the world.

The science-fiction author and science writer Arthur C. Clarke was the first fully to appreciate the impact which communications satellites would have. In October 1945, in the British Journal *Wireless World,* he published a paper entitled 'Extra-terrestrial relays' in which he pointed out how three satellites could be arranged to provide world-wide communications coverage. The key to Clarke's proposal was the geostationary orbit. If a satellite is placed in a circular orbit above the equator at an altitude of 35,800 kilometres it will revolve round the Earth in precisely the same period of time as the Earth takes to spin on its axis. Therefore, despite travelling along at a speed of 11,000 km/hour in its orbit, the satellite will remain vertically above a fixed point on the equator as the two move round in step. To an Earth-based observer the satellite will appear to remain at a fixed point in the sky. Three such satellites, spaced out at equal angles above the equator could link the whole globe (apart from small zones round either pole) for communications purposes.

Clarke was spot-on in this prediction, but even he did not get everything right, for he imagined that communications satellites would have to be large, permanently manned structures, with onboard engineers to maintain the communiccations system. In 1945 no-one could have foreseen the electronics revolution and its miracles of micro-miniaturization.

Global *radio* communication is possible without the aid of satellites, since long-wave radio signals may be reflected from the ionosphere – the layers of ionized (electrically charged) gas in the upper atmosphere – and by this means 'bounced' round the globe. The quality of reception is variable, being prone to hiss, crackles, and fade-outs as changing levels of solar activity cause fluctuations in the reflective properties of the ionosphere. The much shorter waves (higher frequency) of television broadcasts are not reflected in this way, but pass instead straight out through the atmosphere – possibly to the considerable entertainment of any intelligent creatures who may be within a few tens of light-years' range! Television reception occurs along the line of sight, which is why transmitters are normally located as high as possible, and why so many are needed to give national coverage.

With direct transmission to a small number of satellites which receive, amplify and re-transmit the signals, global television coverage becomes a practical reality, and long-range 'radio' communication becomes more reliable and of higher quality. The ultra-high frequencies used today for satellite communications allows many thousands of individual voice channels (telephone 'lines') to be handled by a single satellite, and offer great scope for data transmission.

There are two types of communications satellite – active and passive. A passive satellite has no receiving or transmitting apparatus, but acts as a reflector off which signals may be bounced. The first and most celebrated of these was Echo I, a 30-metre diameter sphere of thin plastic coated with a film of aluminium to make it highly reflective, which was launched and inflated in orbit on 12 August, 1960. Being large and highly reflective, it was readily seen in the night sky and became a familiar sight to amateur satellite watchers, as did its successor, Echo II, in 1964. The Echo satellites were used successfully to reflect radio signals but, of course, such satellites are highly inefficient, for only a minute fraction of the transmitted power is bounced back.

Real progress depended upon active satellites which receive, amplify and re-transmit the broadcasts. The first satellite which had the means to receive and transmit was Atlas–Score, launched on 18 December, 1958, and which broadcast a tape-recorded Christmas message from President Eisenhower to the world. On 1 January, 1961, President Eisenhower declared as a national objective 'the early establishment of a communications satellite system which can be used on a commercial basis', and on 10 July, 1962, *Telstar*, the first privately-financed communications satellite – built by the American Telephone and Telegraph Corporation – was placed in an elliptical orbit, ranging in altitude between 5600 kilometres and 950 kilometres, and having a period of 158 minutes. On 23 July communications history was made when *Telstar* relayed the first live transatlantic television programmes – two twenty-minute broadcasts, one from America to Europe, the other in the opposite direction.

The programmes could not have been significantly longer than 20 minutes for the very good reason that *Telstar* was in a relatively low orbit of short period and did not remain within sight of transmitting and receiving stations for long between rising and setting.

The following month President Kennedy signed into law the Communications Satellite Bill to establish the Communications Satellite Corporation (COMSAT) to take charge of the United States end of future global satellite networks.

The first communications satellite to be placed successfully into a geosynchronous orbit was Syncom II, launched on 26 July, 1963, and stationed above the Atlantic ocean to provide television, teletype, and photo-facsimile communications between the U.S.A. and the African continent, but because its orbit was inclined to the equator by an angle of 38°, it was not truly geostationary; instead it traced out a 'figure of eight' in the sky as it roamed to and fro across the equator between 38° N and 38° S. Next in that series was Syncom III which was stationed over the Pacific and was the first true geostationary satellite. It brought home the potential of the communications satellite with dramatic force by relaying the Tokyo Olympic games of 1964 and allowing them to be viewed by television networks the world over. Today we are so used to receiving live broadcasts and up-to-the-minute television reporting from all over the globe that most of us tend to forget that satellites are involved at all.

The formation of the COMSAT Corporation was followed in 1964 by the founding of the International Telecommunications Satellite Consortium (INTELSAT), with an initial membership of 14 nations, to establish global satellite links. Ten years later the number of participants had increased to 60, and by early 1982, the number of members had swelled to 106. Their first satellite was Intelsat I, better known as 'Early Bird', which was placed into geostationary orbit in April 1965. The 40-kilogram satellite had the capacity to handle 240 telephone channels or one television channel.

Subsequent generations of Intelsats have greatly increased the scope, quality, and capacity of the coverage and, currently, these satellites handle about two-thirds of the global public communications traffic. The first full global television link-up was achieved on 27 June, 1967 with the aid of Intelsat I, and two of the Intelstat II series. Including the first successful launching on 18 December, 1968, a total of five Intelsat III satellites were placed in geostationary orbit up to April 1970, each of them being four times as heavy as and having six times the communications capacity of Intelsat I, and with these a viable global network was established. The Intelsat IV series, launched between 1971 and 1978 increased the capacity yet further to between 4000 and 6,000 telephone channels or 12 colour television channels, per satellite.

The launching of the first of the current generation of Intelsat V satellites came in December 1980. These, the most powerful commercial communications satellites currently in operation weigh just over one tonne and each carries 12,000 voice circuits (telephone channels) and two colour-television channels. Altogether it is expected that 15 Intelsat Vs will be launched, 2 of them being scheduled for launching by Ariane. The next generation Intelsat VIs, with 40,000 voice channels should begin to be launched in 1986.

Other rapid growth areas are domestic communications satellites – for internal communications within individual countries or neighbouring groups of countries and, of course, military communications systems. Domestic communications satellites are particularly valuable to large countries with widely dispersed centres of population. Canada, for example, began such a programme with the first of its 'Telesat' series – named ANIK (an eskimo word meaning 'brother') – launched by NASA in 1972. Internal communications in the United States are handled by COMSAT Corporation's 'Comstars', Western Union's 'Westars' and Radio Corporation of America's 'Satcoms'. Many other nations are developing this kind of capability.

The Soviet Union has been deeply involved in satellite communication, particularly for internal communications and communication within the Eastern bloc. Its 'Molniya' (meaning 'lightning') series, which is continuing still, began with

Molniya 1 in 1965. Because Soviet territory and launch sites are far north of the equator it is a difficult operation to place satellites in geostationary orbits over the equator. The Molniyas follow an ingeniously-selected 12-hour orbit which is highly elliptical with perigee typically at about 300 kilometres altitude and apogee at about 40,000 kilometres. When near apogee the satellite moves very slowly in the sky and is then visible over most of the Soviet Union and can be used for long-range communications for periods of 8 to 9 hours at a time. The time spent near perigee is small because the satellite is then moving very rapidly. Because of the 12-hour period a given satellite will be overhead at the same time each day, and a network of these satellites provides continuous coverage.

Other, more recent, series of Soviet communications satellites have been placed in 24-hour geosynchronous orbits.

Apart from a few experimental satellites, the transmission powers of operational communications satellites are low and their signals must be received on Earth by large ground-station antennæ then relayed through the normal television network to domestic television receivers. The next big step, and the one which will bring the real communications revolution, will be direct transmission from satellites to households. Modest mass-produced antennæ, of about 1-metre diameter, will allow individual households to tune in to a multitude of broadcasts from a wide range of satellites, regardless of national barriers.

The first satellite with the power to achieve this was the NASA-launched ATS–6. Placed into orbit in May 1974 it carried out experimental direct broadcasts to receivers in the U.S.A. and in India by means of its giant 9-metre antenna. Canada, Japan, and the U.S.S.R. are developing systems of this kind, and many other nations and groups are working towards this end. The largest and most powerful currently projected communications satellite is OLYMPUS, currently being built for ESA by British Aerospace; due to be placed in geostationary orbit by Ariane in 1987, it will be distinctive because of the enormous 'wings' of solar panels necessary to supply the power for direct transmission to small antennae.

The scope for the use, and abuse, of this new facility is

Plate 9. OLYMPUS, one of a new generation of high-powered communications satellites being developed for direct television broadcasts to individual homes, and for business communications between commercial organizations. The first version of this 2.3-tonne satellite is due to be launched into geostationary orbit in 1987 by the Ariane launcher. (British Aerospace)

enormous. Once again, Arthur C. Clarke was the first to foresee the full potential of direct transmissions which completely bypass ground stations when he remarked, in 1968, that 'In the long run the communications satellite will be mightier than the ICBM'. It will open up the underdeveloped nations and spread education, it will allow instant global communication, and probably cut down on the amount of physical travelling on which businessmen earn their ulcers today; direct face-to-face communication from office to office or home to home will obviate much of the need for that. An ever-increasing number of television channels will pump yet more entertainment into our homes (What a prospect!)

and the world will shrink yet further. How we shall react to this Brave New World remains to be seen.

Earth Watching

Meteorological satellites have played an increasing role since Vanguard 2, launched on 17 February, 1959, first relayed photographs of cloud cover as seen from space. The first purpose-built weather satellite was the American TIROS–1 (*T*elevision and *I*nfra-*R*ed *O*bservation *S*atellite), placed into a near-circular orbit at an average height of 720 kilometres on 1 April, 1960. With an orbit inclined to the equator by 48° and an orbital period of 99 minutes Tiros was able to range over about half the globe, and in its operational lifetime returned over 22,000 cloud photographs. Later satellites in the series were placed in orbits inclined by 81° which allowed them to scan practically the whole globe; as well as photographing clouds – with ordinary light by day and infrared radiation by night – these satellites also proved their worth in charting the break-up of ice cover and charting routes for shipping.

Two systems were developed for returning the photographic data to the Earth. One system involved taking photographs and storing the data until the satellite passed over the controlling ground station to which the data was transmitted; the other system, installed for the first time on Tiros–8 in December 1963, was Automatic Picture Transmission (APT) whereby pictures were transmitted continuously as the satellite pursued its orbit. Each picture, taken with a wide-angle lens, covered an area 2000 to 3000 kilometres wide, and could be picked up by cheap and simple means by any ground station or amateur enthusiast equipped with a cathode-ray tube or a commercial facsimile receiver. On each fly-by of the satellite a ground station can receive two or three cloud-cover pictures of its own area, and pictures of this kind have become familiar to us all from their use in television weather programmes.

The first regular network of operational meteorological satellites comprised the ESSA (Environmental Science Services Administration) series, commencing in 1966. These satellites were placed in Sun-synchronous orbits. This type of orbit – of great importance to Earth-scanning satellites – is one inclined to the equator by an angle of more than 90° (usually between 97° and 102°, depending on the altitude of the satellite); due to the gravitational effect of the Earth's equatorial bulge, the *orbit* itself slowly rotates around the Earth (i.e. precesses) completing one revolution per year. As a result the plane of the satellite's orbit makes a constant angle with the Sun–Earth line as the Earth moves round the Sun (Fig. 6); and the Earth beneath the satellite's track is always illuminated at a constant angle; i.e. each strip of observations is made at the same 'local time' as the Earth rotates below the satellite's orbit. In this way a uniformly illuminated picture of the whole globe may be assembled.

Low-level satellites, then, photograph a strip of the Earth a few thousand kilometres wide on each revolution, so that over a 24-hour period each part of the globe is viewed once in daylight and once at night. Since many weather systems develop significantly over shorter timescales, a better picture of the changing global weather pattern can be obtained from geosynchronous satellites, each scanning a complete hemisphere from a stationary position. The first geostationary weather satellite was ATS–1 (*A*pplications *T*echnology *S*atellite) which provided the first high-resolution pictures taken from geostationary orbit after its launching in December 1966. This was followed by the SMS (*S*ynchronous *M*eteorological *S*atellite) series.

Sophisticated scanning devices are carried aboard the latest generations of meteorological satellites such as the GOES series (*G*eostationary *O*perational *E*nvironmental *S*atellite) funded by NOAA (*N*ational *O*ceanic and *A*tmospheric *A*dministration), and the Tiros-N (NOAA) series. The instrument carried aboard GOES–4 – launched in September 1980 – and later members of that series is called the Visible Spin-Scan Radiometric Atmospheric Sounder (mercifully abbreviated to VAS). This device detects reflected sunlight in the visible part of the sprectrum, and also measures infrared radiation at twelve different wavebands, emanating from different levels in the atmosphere, so obtaining information from different heights in the atmosphere, from ground level upwards.

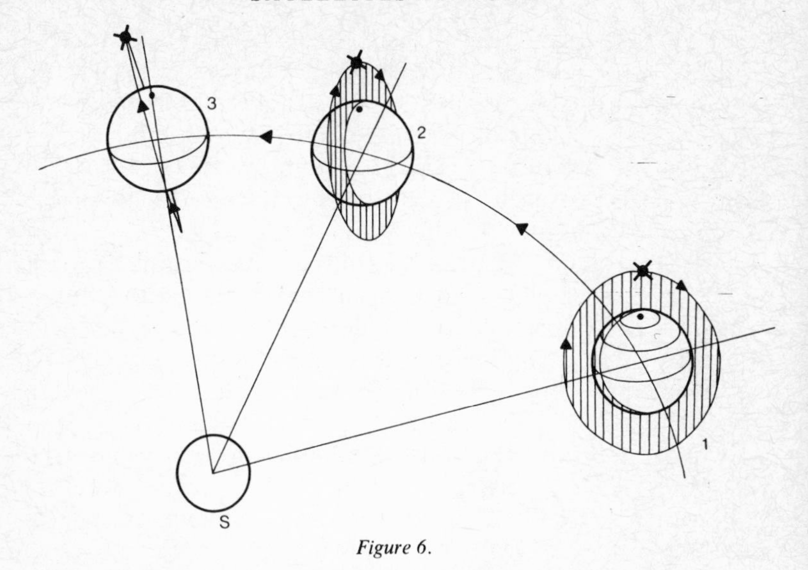

Figure 6.

Sun-synchronous orbit. As the Earth moves round the Sun, from (1) to (2) to (3), etc., so the plane of the satellite's orbit (shaded) rotates round the Earth in such a way that it maintains a constant angle to the Sun–Earth line (see text).

By means of this instrument the familiar pictures of the Earth's surface and cloud cover are obtained, together with sea surface temperature data, and a vertical scan through the atmosphere giving details of temperatures and of the amounts, distribution, and movement of water and water vapour at different heights. This sophisticated, 20 million-dollar satellite also collects and relays back to Earth environmental data from more than 1500 instrument platforms remotely scattered on land, at sea, and in the air (on balloons and aircraft). It also makes measurements of solar activity and the Earth's magnetic field, so providing a very comprehensive environmental coverage.

These and other classes of satellite have progressed a long way from simply taking cloud pictures. They provide day and night coverage of clouds, oceans, ocean currents, ice distribution, winds, sea state, wave heights, and can even distinguish between rain, hail, and snow clouds, as well as linking all these phenomena to solar radiation and the geomagnetic field. The more recent NOAA satellites provide sea temperature data – of interest to fishermen looking for the best fishing grounds – to an accuracy of about 1 °C, and even carry equipment capable of picking up distress beacons. Ocean current and sea state data allow oil tankers and other vessels to make the best use of current to achieve considerable fuel savings (figures of as high as 20–40 per cent have been reported, and 10 per cent is quite readily attainable).

The Soviet Union, too, has established an elaborate network of METEOR satellites working in conjunction with ground station and balloon data, to provide a comprehensive meteorological service giving storm warnings, details of cloud cover, ice conditions, general weather forecasts, and details of winds and current which, among other things, allows Soviet shipping to follow the best and most economical routes. Close co-operation exists between Soviet and American weather departments, the interchange of information helping to build up a global view of weather and climate.

Following a series of ten purpose-built Cosmos satellites, which began in 1964, with the aim of developing an operational satellite, Meteor 1 was launched on 26 March, 1969. To date there have been 36 launches in the series. Like all the early American meteorological satellites the Meteor series flew fairly low orbits.

Other nations, too, are involved in meteorological/environmental monitoring satellites. For example, the European Space Agency has launched several in the Meteosat series since 1977.

Another fruitful and expanding area of satellite applications is the study and monitoring of Earth resources. By imaging the Earth's surface at different wavelengths, satellites can provide useful information on agriculture, forestry, land and sea resources, and the geological data which they supply can provide guidance to the location of mineral resources.

Best known of this type of satellite is the American LANDSAT series (originally known as Earth Resources Technology Satellites – ERTS), the first of which was launched on 23 July, 1972, and which collect data to be relayed directly to designated ground stations. Landsat's instrumentation makes images of the same area of terrain through filters transmitting light at a number of different wavebands in the visible and near-infrared part of the spectrum. By combining the details visible at the different wavebands an extraordinary amount of information can be revealed. For example, vegetation shows up as relatively bright in the yellow-green band, dull in the red, and bright in the infrared; by comparing the brightness of a region at the different wavebands, the amount of vegetation cover can be obtained and, furthermore, from the precise characteristics of the reflected light a good estimate of the type of vegetation can be made and the state of health of crops can be investigated.

The pictures obtained at different wavebands are often presented in rather striking form by dying each image with a different colour and then combining three or four to form a 'false colour' image in which the salient features of all of them stand out clearly. These images are both informative and attractive.*

Landsat photographs have allowed geologists, geographers, biologists, anthropologists, oceanographers, cartographers, civil engineers, and others to carry out survey work – particularly in remote and inaccessible areas – more comprehensively and much more rapidly than ever before. In some areas of the globe, aerial reconnaissance photography has accomplished surveys which would have been lengthy, difficult, dangerous, or even impossible to accomplish from ground level, but only a fraction of our planet has been adequately covered in this way; Landsat and its cousins are opening up the whole globe to this kind of investigation. From charting the spread of crop disease or detecting forest fires, to mapping uncharted areas of the globe or plotting routes for roads and railways, satellites of this kind have so much to offer.

Landsat data is made available directly to designated ground stations, and – to date – ground stations have been established in the United States, Canada, Brazil, Italy, Sweden, Japan, India, Australia, Argentina, and South Africa, and it is expected that stations will soon come into operation in the People's Republic of China, Thailand, and Indonesia.

* *Earthwatch* by Charles Sheffield (Vice President of Earth Satellite Corporation), published 1980 by Sidgwick and Jackson Ltd., London, contains magnificent examples of these images.

Navigational aids

Satellites have also been developed as an aid to navigation – an idea which was first put forward by the American writer Edward Everett Hale in a story called 'The Brick Moon', published in 1869-70. The first step towards establishing a network of satellites for this purpose was taken in April 1960 with the launching of Transit 1B, the first of a series of satellites in low Earth orbits which allowed ships and aircraft to determine their position by means of the 'Doppler effect'. If a source of radio waves is approaching an observer the frequency of the signal is increased (compared to the frequency released from the transmitter) by an amount proportional to the velocity of approach; likewise the frequency of a receding source is decreased in proportion to its velocity of recession. In a similar fashion the pitch of an approaching sound (a police siren, say) is high and then immediately drops as the source passes by and begins to recede.

If both the orbit and velocity of the satellite is known, measurements of frequency changes in the signal made at regular intervals as the satellite approaches then recedes (its frequency will be higher than nominal on approach, nominal at closest approach, and lower than nominal as it recedes) allow the navigator to plot a series of two or three position lines (lines on the Earth's surface on which he must be located for the satellite signal frequency to change in the observed manner), and where these lines cross is his position on the globe.

Because the satellites are at low altitudes, they move rapidly across the sky and a given satellite will allow the navigator only a few 'fixes' per day. With the current network of satellites, the average interval between suitable fixes is about 1.5 hours. The current 'transit' or 'Navsat' system consists of about five operational satellites following polar orbits with periods of about 108 minutes; their orbits form a kind of grid below which the Earth rotates. Precise details of each satellite's orbit are transmitted from the satellite and these are fed automatically into the ship's or aircraft's satellite navigational microprocessor. Provided certain basic information has been keyed in, *Satnav* systems will readily give the position of the observers's craft in the form of a digital display of latitude and longitude to an accuracy of about 100 metres. Continuous-position fixing is not yet possible, but this facility should become available by the late 1980s with the more sophisticated Navstar Global Positioning System which should be coming into operation around that time, and which will offer positional accuracies of about 10 metres.

Originally developed for the U.S. Navy, *Satnav* has now become a world-wide service. At first *Satnav* equipment was to be found only on the most sophisticated ships and aircraft, but it has now beome an integral part of the electronics revolution which has struck small boat navigators. In 1966, when the equipment first became commercially available, a *Satnav* receiving system cost about £30,000; today it is possible for a yachtsman to buy a compact system for under £1,000 which is capable of yielding his position to within a tenth of a nautical mile. Such is the measure of progress.

Military machinations

That satellites should have important military applications is self evident. Military satellites are used to provide communications, navigational data, meteorological information and, above all, to provide an unrivalled reconnaissance capability, transcending all national boundaries. Both the U.S.A. and the U.S.S.R. have devoted considerable resources to such systems, and a considerable fraction of all satellite launchings by both superpowers fall into the category of reconnaissance or 'spy' satellites.

United States activity in this field began with the Discoverer series of satellites which were basically Agena upper-stage vehicles equipped with cameras and which tested the technique of returning film packages from space. The first successful retrieval of a re-entry capsule was achieved in August, 1960, when the Discoverer 13 capsule was fished out of the Pacific Ocean. Subsequently, re-entry capsules have been recovered in the air by catching them in large nets towed behind aircraft. (Stories about 'the one that got away' probably don't go down too well in this context!) During 1961 about a dozen Discoverer missions were flown, and it is believed that they made the major strategically important discovery that, contrary to what had been rumoured in the West, the Soviet

Union was not undertaking a major build-up in its deployment of ICBMs. This finding almost certainly saved the United States from the expense of a major escalation of its own missile programme.

The best quality spy-photographs are obtained by satellites which fly in low orbits and dip down into the atmosphere over selected targets, photographing directly on to film and then returning that film to Earth. The Soviet Union uses craft similar to unmanned versions of the Vostok spacecraft for this purpose, but flying under the heading of Cosmos. The first such vehicle to be recovered was Cosmos 4, in April 1962.

In the U.S.A. Samos (Satellite and Missile Observations System), launched into polar orbits – like most American military satellites, from the Vandenberg Air Force Base, California – were used to seek out new weapons deployments. The Midas (Missile Defence Alarms System) satellites carried infrared detectors to pick up the heat released as ICBMs are launched. The Vela satellites were placed in very high orbits (over 100,000 kilometres) to look for nuclear detonations on the Earth or in space. The first of the Vela series was launched in 1963 to watch for violations of the nuclear test ban treaty. One mystery which does not seem to have been resolved is the nature of a flash of light detected near Antarctica by a Vela satellite on 22 September, 1979. Suggestions were made that this had been a clandestine nuclear test carried out possibly by South Africa, but this interpretation has been strenuously denied and the question remains – publicly at least – unanswered.

Among the more recent generations of spy satellites have been the American 'Big Bird' devices, weighing nearly 14 tonnes, which – since 1971, have been launched into polar orbits by Titan III D launch vehicles. Their high resolution cameras reputedly can resolve details only 0.3 metres across. Clearly satellites of this capacity can reveal not only the movement of military vehicles but also objects as small as tables, chairs, and people – which is a sobering thought if you happen to be reading this paragraph in the open air.

Sinister though they may be, spy satellites have served a very useful role over the past two decades and, with justification, may be said to have preserved the peace of the world by preventing the possibility of clandestine weapons deployment, and providing the key to the effective monitoring of nuclear test ban treaties and arms limitation agreements without the necessity for on-site ground inspections which previously had been the major stumbling block in the way of any such agreements. We must be grateful to them for that.

A more disturbing development of late is the possibility of satellites being used in an offensive role as 'killer' satellites to destroy or disable opponents' satellites. The Soviet Union in particular has developed a class of satellite which matches orbit with its target, moves alongside, then explodes, destroying itself and its target; the first successful trial of this system took place in October 1968 when Cosmos 249 sidled alongside Cosmos 248, and exploded.

Another anti-satellite system known to be under development is the use of high-powered lasers to 'blind' a target satellite, so putting its sensors out of action. For the moment, satellites are defenceless against such attacks; the cost of developing anti-anti-satellite systems would be horrendous, but this may become an inevitable consequence of any stepping up of the offensive role of satellites (see Chapter 12).

Over the past 25 years, the satellite has assumed a multitude of roles and has exerted its influence on many aspects of everyday life. It is changing the face of the world, whether we like it or not; it is already bringing enormous benefits but, like all such developments it has its drawbacks, too. The myriad of new prospects opened up by the satellite communications and surveillance revolutions have so much to offer, but careful safeguards will be required to prevent their abuse. But as national barriers are transcended, the satellite may become instrumental eventually in unifying a fragmented and suspicious world. If so, that will be its greatest achievement. Much has happened since October 1957, but the full impact of the satellite era will be felt in the next 25 years.

8

Man in Space

On the morning of 12 April, 1961, an A–1 class launch vehicle stood on the launch pad at the 'Baikonur Cosmodrome', near Tyuratam in the Soviet province of Kazakhstan. At 09.07 hours Moscow Time the pad was lit by the glare and shaken by the roar as the twenty main engines and twelve steerable vernier motors burst into life. As the thrust built up to over 500 tonnes, the vehicle – standing 38 metres tall – slowly at first, but with increasing acceleration surged upwards, reaching for space. Just a few minutes later its payload, the 4.7 tonne spacecraft Vostok 1 was placed in orbit . . . aboard was Major Yuri Gagarin, *the first man in space*.

Gagarin's spacecraft made one complete circuit round the Earth reaching a maximum altitude of 327 kilometres, and its re-entry module landed safely 108 minutes after lift-off. The capsule touched down with the aid of parachutes, but Gagarin himself was flung from the craft by means of an ejector seat at an altitude of about seven kilometres and made the final descent under his own parachute. At that stage the Soviet mission planners – who, unlike the Americans, had little option but to make a touchdown on hard land rather on the slightly more yielding ocean – were not prepared to risk landing a man directly in his space capsule.

News of Gagarin's flight took the world by storm. Instantly, his name was on everyone's lips, and it remains a household word today. Indeed, for as long as the human race survives, the name of the first man to leave this planet will surely be recalled. Although the flight was controlled entirely automatically, and Gagarin would have been permitted to

Plate 10. Yuri Gagarin (left), first man to fly in space with Sergei Korolev, Chief Designer of the Vostok spaceship and architect of the early Soviet space programme. (Novosti Press Agency)

take over command only in the gravest emergency, it was a courageous and dramatic flight into the unknown. He suffered no ill-effects, and at once laid to rest many of the fears which had earlier been expressed about possible harmful effects which spaceflight might produce – the hitherto unknown effects of radiation, weightlessness, and the crushing accelerations which – although simulated by training machines, had never before been experienced 'for real'.

Banner headlines leapt from newspapers the world over, unanimous in saluting what the London *Evening News* called 'the greatest feat of all'. Gagarin was ideally suited to the task in every respect – not only as a trained cosmonaut in the peak of his physical fitness at the age of 27 – but with the looks and personality to become an ideal ambassador of space

Plate 11. Mock-up of the Vostok spacecraft in which Yuri Gagarin made his historic first manned spaceflight on 12 April, 1961. (Novosti Press Agency)

exploration. Alongside its front-page report of the flight, the London *Evening Standard* carried a picture headlined 'Spaceman No 1' and captioned 'Yuri Alexeyevich Gagarin . . . a good honest smile . . . fine intelligent eyes.' In the years which followed – up until his tragic death in an air crash in 1968 – he made many trips abroad to describe his experiences to an admiring world.

Rumours of earlier (and later) cosmonauts who had been lost in space prior to and subsequent to Gagarin's flight were almost certainly completely and utterly false, although they enjoyed a good deal of publicity at the time. In fact, the flight of Vostok had been preceded by a quite elaborate series of unmanned test flights of prototype spacecraft dating back, in a sense, to the flight of Laika in Sputnik 2. Beginning with Sputnik 4 (otherwise known as Spaceship 1) launched on 15 May, 1960 a series of five Vostok-type craft were flown.

The planned recovery of Spaceship 1 failed because the spacecraft was inclined at the wrong attitude when the retro-motor fired; instead of re-entering the atmosphere, the spacecraft entered a higher orbit. Spaceship 2 (Sputnik 5), launched on 19 August of that year successfully returned to Earth its crew of two dogs – Strelka and Belka – but the next two dogs to go into space were less fortunate when Spaceship 3 burned up in the atmosphere. The successful orbital flights and safe return of dogs aboard the fourth and fifth Vostok precursors set the seal on the test programme and set the stage for Gagarin's epoch-making flight.

Vostok 1 was followed by Vostok 2, launched on 6 August, 1961, and aboard which Herman Titov completed 17 orbits in a flight which lasted 25 hours 18 minutes.

Beaten to the post once again by the Russians, the mood of America was such that a spectacular space 'first' for the United States was required to restore battered American self-esteem and renew confidence in America's technological prowess. Something dramatic was called for. A new, youthful President sought for a bold new target for American ingenuity. In a memorandum to Vice-President Lyndon B. Johnson, dated 20 April, 1961. President John F. Kennedy wrote:

> 'Do we have a chance of beating the Soviets by putting a laboratory in space, or by a trip to the Moon, or by a rocket to go to the moon and back with a man. Is there any other space program which promises dramatic results in which we could win?'

Just five weeks later President Kennedy set in motion the greatest space programme of them all, the Apollo Project, with this declaration of 25 May, 1961:

> 'I believe this nation should commit itself to achieving the goal, before this decade is out, of landing a man on the moon and returning him safely to Earth. No single space project in this period will be so impressive to mankind, or more important for the long-range exploration of space, and none will be more difficult or expensive to accomplish.'

Just six weeks after Gagarin's triumphant spaceflight, the gauntlet was down and the space race was on in earnest.

The United States had formally initiated its manned space programme with Project Mercury – officially sanctioned by NASA on 7 October, 1958, just six days after the space agency had started business. The aim of the project was to send men into space to investigate their capabilities and reactions. With the 'Kennedy declaration' the project became the key first stage in developing the means to permit a man for the first time to set foot on another world and return safely to the Earth. The selection of a suitable launch vehicle, the design of the spacecraft, the selection and training of the crews were major tasks which the project team took on board. The seven astronauts who were to fly the missions were announced on 9 April, 1959. The first flights were to be sub-orbital 'hops' – up into space and back down without attempting to complete an orbit; later flights would aim for full orbital missions.

The launch vehicle selected for the sub-orbital missions was the redoubtable Redstone missile; orbital flights were to be launched by the U.S. Air Force's revolutionary new Atlas ICBM when it became available. The modest thrust of the Redstone – a mere 35 tonnes compared to the 500 tonnes of the Soviet A–1 – set severe limitations on the weight of the manned spacecraft, so that the term 'capsule' became particularly appropriate. Designed and built by the McDonnell Aircraft Corporation, the Mercury spacecraft was a triumph of compact engineering.

The conical, almost bell-shaped spacecraft measured 2.9 metres in height with a maximum diameter across the base of 1.9 metres. Inside, there was just enough room for the astronaut lying on his contoured couch and for the necessary masses of instrumentation. The broad blunt base of the spacecraft was covered by a heat shield made of ablative material which would melt and carry away excess heat as the craft made its fiery re-entry – base first – into the atmosphere. For the launch a small solid-fuelled rocket was attached to the top of the spacecraft to act as a last-ditch escape device which would hurl the spacecraft clear of the launch vehicle in the event of some disaster in the initial stages of the flight.

Descent through the atmosphere would be slowed by friction, then by a conical drogue parachute opening at about 6 kilometres' altitude, and the final descent to splashdown in the ocean would be made with the aid of a 19-metre diameter parachute opening at an altitude of about 3 kilometres. The launch site was to be Cape Canaveral in Florida.

The first unmanned trials of Mercury–Redstone, and of Mercury–Atlas were unsuccessful, but a satisfactory sub-orbital, unmanned test flight was achieved on 19 December, 1960. The next flight, MR–2, on 31 January, 1961, carried the chimpanzee 'Ham' to an altitude of 253 kilometres (68 kilometres higher than planned) through a period of weightlessness lasting 6 minutes, 40 seconds to a splashdown about 212 kilometres off target. The remarkably tolerant chimp was picked up unharmed.

On 5 May, 1961, Commander Alan Shepard entered his Mercury capsule – which he had named 'Freedom 7', and at 09.34 Eastern Daylight Time the Redstone blasted off, its engine firing for 142 seconds to take Shepard's capsule to a maximum speed of 8,260 km/hour after which Freedom 7 coasted to an altitude of 188 kilometres before falling back to splashdown 486 kilometres down range after a flight lasting 15 minutes 22 seconds and including just over 5 minutes of weightlessness. Although eclipsed in grandeur by the flight of Vostok 1, the successful Freedom 7 mission – made in the full glare of world-wide publicity and television coverage – aroused jubilation in the American camp.

Two months later Air Force Captain Virgil 'Gus' Grissom made a similar sub-orbital flight in the capsule 'Liberty Bell 7'. Grissom's worst experience came after splashdown when the hatch cover blew off and Liberty Bell began to fill with water. Grissom swam away and was safely picked up, but despite the best efforts of the pick-up crew, the spacecraft sank to the bottom of the Atlantic.

After a chimpanzee test flight of Mercury–Atlas on 29 November, 1961, the way was clear for the first manned orbital flight of the series. After about four weeks of postponements John Glenn entered the capsule 'Friendship 7' atop the Atlas vehicle. Holds in the countdown kept him there for 3 hours 44 minutes before lift-off finally took place at 09.47 on 20

February, 1962. Just five minutes later the sustainer engine of the Atlas shut down and the 1.4 tonne capsule was inserted into a near perfect orbit of apogee 261 kilometres and perigee 161 kilometres. Ten months after Gagarin's flight, John Glenn became the first American in orbit and the first to witness the grandeur of repeated sunrises and sunsets from orbit.

One mystery associated with this flight, and widely reported at the time, was Glenn's sighting of thousands of little specks of light floating round the capsule around the times of sunrise, which occurred every 1½ hours as he circled three times round the Earth. These 'fireflies' as he called them, were also seen by later astronauts and may have been due to paint flaking off the hull of the spacecraft.

Problems arose with sticking valves in the attitude control thrusters, forcing Glenn to take over manual control, and a major alarm arose when telemetry data indicated to ground control that the vital heat shield was no longer locked in position. It was decided to keep the retro-rocket package in place during re-entry (it was normally jettisoned after use) in the hope that it would hold the heat shield in position as the spacecraft plummeted back through the atmosphere. All went well, re-entry was safely accomplished, and 'Friendship 7' splashed down in the Atlantic Ocean about 65 kilometres short of the predicted target area. Seventeen minutes later the U.S. Naval destroyer 'Noa' cruised alongside and Glenn's capsule was hoisted smartly aboard. The only minor injury suffered by Glenn was a cut on the knuckles received when he struck the plunger to blow off the hatch.

With Glenn's safe return a hundred million television viewers breathed a sigh of relief. Glenn became a national hero. Much had been done to restore national pride in American technology.

Three further Mercury-Atlas flights ensued, the series concluding with L. Gordon Cooper's mission, launched on 15 May, 1963, which completed 22 orbits and raised the American space endurance record to 34 hours. In all the project had cost just over $400 million inclusive of the tracking and data acquisition costs; it achieved all its principal objectives and set the stage for the next development on the road to the Moon.

Despite its undeniable success, Project Mercury was overshadowed by the continuing triumphs of the Vostok series. Vostok 3, launched on 11 August, 1962 with cosmonaut Andrian Nikolayev aboard, was joined in orbit the next day by Vostok 4, piloted by Pavel Popovich. For the first time two manned spacecraft were in orbit simultaneously. The two Vostoks passed within 6.5 kilometres of each other, but they were not capable of being manœuvred by the cosmonauts on board. The two Vostoks touched down within a few minutes of each other on 15 August, Vostok 3 setting a space endurance record of 64 orbits and over 94 hours in space.

The Soviet space programme achieved a new coup on 16 June, 1963, when Vostok 6 entered orbit with the first – and so far the only – woman to go into space – Valentina Tereshkova. Feminist groups in the West pointed to this as an example of how women are treated equally in the Soviet Union, and roundly castigated NASA for not including women in its astronaut training programme (this attitude, at last, is beginning to change), but the fact remains that since that date no Soviet woman has set foot in a spacecraft. Vostok 6 joined Vostok 5 which had entered orbit two days earlier and which went on to set a record of 81 orbits and 119 hours in space which was not to be surpassed for more than two years.

Vostok was followed by Voskhod, the first spacecraft to carry a three-man crew into space; on board were Vladimir Komarov, Konstantin Feoktistov, and Boris Yegorov – the first trained doctor to go into space. The mission, which began on 12 October, 1964, lasted for 24 hours, and all three cosmonauts returned safely to Earth. Subsequent analysis has revealed that Voskhod was not – as had at first been suggested – a new purpose-built three-man craft but was instead a stripped-out Vostok, with its ejector seat removed, and three couches squeezed in, giving sufficient room for three cosmonauts without spacesuits. A retro-rocket was fitted to allow the crew to land in the spacecraft rather than ejecting and descending with personal parachutes (there was not sufficient room in the spacecraft for that provision, in any case). It is quite widely believed that the major purpose of this mission was to place a three-man Soviet spacecraft into orbit before the first flight of the American two-man Gemini spacecraft which at that time was undergoing unmanned test flights.

Voskhod 2 entered orbit on 18 March, 1965, with a crew of two – Pavel Belyayev and Alexei Leonov. Although it carried only two cosmonauts, Voskhod 2 was heavier than its predecessor, having an all-up weight of 5.7 tonnes, the extra weight being due to the inclusion of a new feature, an airlock which would allow access to space without having to depressurize the cabin of the spacecraft.

At 500 kilometres above the Black Sea, Leonov stepped out of the airlock and pushed away from his spacecraft, drifting to the end of the 5-metre tether which attached him to his craft and repesented his only hope of getting back. A television camera mounted outside the spacecraft captured for the television audience below this historic event when, for the first time ever, a man floated in space, sustained against his hostile environment only by his own space suit. The ten-minute 'spacewalk' proved to be exhausting, and Leonov experienced considerable difficulty in getting back into the airlock, mainly because his suit had 'ballooned' outwards while he was in space and getting into the airlock was a tight squeeze. Astronauts clad in restrictive space suits, with no ready means of support or orientation, have consistently found that 'spacewalking' – or EVA (*e*xtra-*v*ehicular *a*ctivity) is in practice anything but the relaxed 'floating around' which had been suggested in fiction.

Belyayev and Leonov had further trials ahead. On the sixteenth orbit the automatic re-entry system failed and on the following orbit, for the first time in the Soviet space programme, the manual back-up system had to be used. Belyayev guided the craft safely down to Earth, but the touchdown took place in deep snow and dense forest in the Ural mountains about 2000 kilometres north-east of the planned landing site. They spent an unhappy night huddled in the capsule in freezing conditions while marauding wolves tried to force an entry into the capsule, and were not rescued until the morning following the landing.

Despite the hitches, Leonov's daring venture into the unknown fired the imagination of the world and chalked up another 'first' for the Soviet space programme. Two years were to pass before the next Russian cosmonaut flew in space – and that voyage was to be made in a totally new spacecraft – Soyuz. In the meantime the American Gemini programme took over centre stage.

The Gemini spacecraft was larger, heavier, and appreciably more spacious than its Mercury predecessor, with a weight of 3.7 tonnes, a length of 5.6 metres and a diameter across the base of 3.05 metres. It consisted of two principal components – the re-entry module which housed the astronauts, and the 'adapter module' which contained instruments, life-support systems, power supplies, and so on. Instead of there being a launch escape tower, each astronaut sat in an ejector seat which could be used for escape not only at launch but at various stages of the flight. The two-man Gemini craft carried an onboard computer which, together with elaborate thruster systems, allowed the crews to exercise a wide measure of control and really 'fly' the spacecraft.

The launch vehicle selected was the Titan II – a two-stage vehicle adapted from the U.S. Air Force's Titan ICBM. More powerful than the Atlas, it had other advantages, too. It was a highly reliable vehicle, simpler (unlike the Atlas, it did not need to be pressurized to prevent its buckling), and it had the benefit of using storable 'hypergolic' propellants (i.e., fuel and oxidizer burn spontaneously when brought into contact.) The propellants did not require elaborate cold storage and handling facilities (and did not need to be off-loaded if a long delay occurred in the countdown); because the propellants were hypergolic no ignition system was required. The lift-off thrust of 195 tonnes was still only about 40 per cent of that of the redoubtable Soviet A–1, but its payload to orbit was comparable to that of the heavier Soviet vehicle.

The Gemini programme was to play a crucial role in developing the techniques necessary for the later Apollo lunar missions. For reasons which will be discussed in the next chapter, reliable precision techniques of orbital rendezvous and docking between two spacecraft were fundamental to the success of the Apollo project, and one of the primary aims of the Gemini series was to perfect this ability. Another was to test man's ability to survive spaceflights of up to two weeks' duration, and his ability to cope with the extended EVA which would be required when the lunar landings were made.

The first two Gemini missions were unmanned test flights, then – on 23 March, 1965, just five days after Leonov's epoch-making spacewalk – Gemini 3, commanded by Virgil Grissom and crewed by John Young blasted into space to make a near-perfect three-orbit mission. During the flight the spacecraft's thrusters were fired to change the orbit from elliptical to near circular. This was an important 'first' as it demonstrated the Gemini craft's capability for orbital manœuvring. On a lighter note, a post-flight furore arose when it was discovered that John Young had smuggled aboard a genuine, non-regulation, corned-beef sandwich which had been partly eaten in orbit!

Gemini 4 launched on 3 June of that same year, with James McDivitt and Edward White aboard, made 62 orbits, extending the American space endurance record to four days. Unsuccessful attempts were made to rendezvous with the launch vehicle in the early stages of the flight, but the event which will always be associated with this mission was Ed White's EVA, a 'spacewalk' of about twice the duration of Leonov's. During the EVA White used a hand-held thruster – a gas gun – as a means of personal propulsion (just like the early comic strip space adventurers!). When ordered back into the spacecraft, White did so with reluctance, entranced as he was by the magnificence of the spectacle around him; the most difficult operation proved to be closing the hatch, a task which left both astronauts in a state of exhaustion.

Gemini 5 made an eight-day flight in August 1965 and so, for the first time, the space endurance record passed from the Soviet Union to the United States. Frank Borman and James Lovell on Gemini 7 clocked up the longest spaceflight of the sixties with a 206 orbit, 14-day mission in December 1965. While in orbit they were joined by Gemini 6, with astronauts Schirra and Stafford aboard; they manœuvred their craft to within 2 metres of Gemini 7, so achieving the first true orbital rendezvous. The degree of control achieved was gratifying, allowing the craft to approach and back off at will.

Neil Armstrong and David Scott took matters a crucial stage further when they docked Gemini 8 with an unmanned Agena target vehicle on 16 March, 1966. When the astro-

nauts started to use the Agena's attitude control system to manœuvre the linked combinations, the Gemini–Agena started to roll. The crew disengaged and backed off, but immediately the problem became much worse, the spacecraft rolling over and over at a rate of one revolution per second. The fault proved to be a manœuvring thruster which had stuck on. By dint of using the re-entry control thrusters the spacecraft was stabilized, but it then became imperative to return immediately to Earth, and a dramatic splashdown was made in the Pacific rather than in the Atlantic. Skill, ingenuity, and training had triumphed over potential disaster in this the first major space emergency.

Geminis 9 to 12 continued to improve upon, extend, and diversify the vital manœuvring techniques, and gained further experience in EVA and in operations such as docking. When Gemini 12, with Jim Lovell and Edwin Aldrin aboard, splashed down on 15 November, 1966 – just 4.8 kilometres

Plate 13. First rendezvous in space by Gemini 6 and 7 on 15 December, 1965. (NASA)

from the waiting carrier U.S.S. *Wasp* – it rang down the final curtain on a series of ten manned and two unmanned missions which had achieved everything of significance which it had set out to do, and more besides. An impressive array of 'firsts' were notched up – first rendezvous; first docking, first orbital manœuvres, first orbital manœuvres by linked spacecraft, and the first transfer to a high altitude orbit which took men into the radiation belts; the first long-duration EVAs (Aldrin clocked up 5½ hours in total on Gemini 12), controlled precision re-entry and landings, and, of course, a space endurance record which was to stand for five years. Ten missions in twenty months had demonstrated man's ability to work in space and carry out all the manœuvres essential to the Apollo missions.

Furthermore, the Gemini programme marked the stage at which the American space programme, which had for so long trailed behind that of the Soviet Union, overhauled and pulled firmly ahead of that of its rival in what was still clearly seen at that time to be very much a 'space race'.

The way ahead seemed clear for Project Apollo. Unmanned prototypes of the three-man spacecraft had successfully been orbited and returned to Earth; the Gemini missions had shown that the astronauts could do all that was required of them. But on 27 January, 1967, tragedy struck. On a ground test of the spacecraft an electrical fault started a raging fire in the pure oxygen atmosphere of the Command Module. The fire gained a hold so rapidly, that – despite frantic efforts by the technicians outside and the astronauts inside – the hatches could not be opened in time and the three astronauts, Virgil Grissom (who nearly drowned on the second Mercury mission), Ed White (who made the first American 'spacewalk') and Roger Chaffee (who had not been into space) were burned to death.

It was a horrific incident which led to modifications in design and to the abandonment of pure oxygen atmospheres in American spacecraft. But the damage could not be undone. Three men were dead, and the first manned Apollo launching had been set back by well over a year.

In the meantime, the Soviet Union had been pressing ahead with the development of a new generation of manned spacecraft, the Soyuz (of which more will be said in Chapter 10). The first manned flight of the new spacecraft – Soyuz 1 – took place on 23 April, 1967, with Colonel Vladimir Komarov (commander of Voskhod 1) aboard. It seems likely that a problem with the spacecraft prompted its early return during the eighteenth orbit. Tragically, the parachute lines became entangled and the re-entry module plummeted to Earth, crashing, and killing Komarov outright. News of the tragedy – the first death to result directly from a manned spaceflight – stunned a world aware of the risks involved in manned spaceflight, but lulled into a sense of security by the successes of the previous six years.

From the launchpad fire of Apollo more than twenty months

were to elapse before Americans again flew in space, and from the Komarov disaster eighteen months were to elapse before another Soviet cosmonaut went into orbit.

The first heady era of manned spaceflight was over. A new one was soon to begin.

9

'One Small Step. . .'

On 20 July, 1969, at 22.56 hours EDT (02.56 GMT of 21 July) Neil Armstrong, commander of the Apollo 11 mission, stepped gingerly from the bottom step of the ladder attached to the lunar module 'Eagle' and placed his boot in the lunar dust. In so doing he opened up a new era in Man's evolution, and wrote his name for ever in the annals of human endeavour, by becoming the first man to set foot on another world – the Moon. On live television the world watched, anxious and fascinated, as Armstrong attained a goal which men had dreamed of for millenia and uttered that slightly indistinct remark 'That's one small step for a man, one giant leap for mankind.'

The successful landing of Apollo 11 and the safe return of her crew marked the culmination of a decade of sustained technological enterprise on a hitherto unprecedented scale. Just eight years and two months after the Kennedy declaration on 25 May, 1961, America had achieved the goal which he set but, sadly, John F. Kennedy was not alive to savour that success, having perished by an assassin's hand in 1963.

That Project Apollo owed its existence to a political decision to demonstrate the technological superiority of the United States to a world deeply impressed by the dramatic achievements of the Soviet space programme matters little in the long term. If the human race survives to spread through the Solar System and, perhaps, beyond, Project Apollo will be remembered as a key step in shedding the reins of Mother Earth when the political wranglings of the East-West conflict have long since been confined to the dusty recesses of academic archives. And Neil Armstrong and Edwin Aldrin will be remembered as the men who took that step.

Ironically, the expected political capital did not really materialize. The world was impressed by a magnificent technological achievement and saluted the skill and courage of the men who undertook the missions, but the global political situation had moved on; America was deeply embroiled in the costly Vietnam War, social problems at home were held up in contrast to the resources stated by opponents of Apollo to have been 'squandered' on the project. But such carping with the 'wisdom' of hindsight is easy, trite, and pointless, and ignores the atmosphere which existed at the time when Project Apollo was born.

To those of us whose imaginations had ever been fired by thoughts of manned space travel – to witness a voyage to the Moon in our own lifetimes was the most exciting and fulfilling prospect. Apollo opened men's minds to real possibilities which exceeded the best of science fiction. It offered challenge, and a sense of purpose in a strife-torn world. Whether the prime motivation is seen by the judgement of history to have been Man's insatiable desire to extend the frontiers of exploration, or the political need for a nation to re-establish its technological pre-eminence, the fact which remains is that men went to the Moon in 1969 and returned safe to Earth, and nothing can change that fact.

Prior to the Kennedy declaration a manned landing on the Moon had been seen as a much longer-term operation, involving the development of a truly enormous rocket, 'Nova' to permit a direct flight from the surface of the Earth to the Surface of the Moon, and a direct return. To achieve a lunar landing before 1970 another technique was adopted which was much more economical in terms of energy, although complex in its operation: *lunar orbital rendezvous*. In this scheme, a spacecraft would travel from Earth into orbit round the Moon and a smaller, ferry craft would separate from it to take the crew to the lunar surface. After completion of the landing mission, the ferry would return the crew to the orbiting spacecraft which would then return to Earth. The importance of the precision rendezvous and docking operations, perfected by Gemini, is self-evident.

Even using this technique, nearly 50 tonnes of spacecraft would need to be placed in orbit round the Moon, and to achieve this demanded a new launch vehicle of far greater power than anything in existence at the time Project Apollo was conceived. As it happened, work had already begun on the development of a powerful launch vehicle for civilian space missions – the first major American launch vehicle to be developed without any immediately obvious military applications. The concept which had originated with Wernher von Braun and his team, was for a vehicle using a cluster of existing rocket motors to generate a total thrust of some 700 tonnes, and the development work was officially sanctioned in August, 1958. Unofficially the clustered launcher was known as Juno V, but in October of that year von Braun proposed the name 'Saturn', and this was adopted; it fitted in with the American convention of naming launch vehicles after Roman or other mythological gods, and as a follow-on from the 'Jupiter-C' and 'Juno', it seemed most appropriate.

The first series type, designated Saturn I, made a maiden flight on 27 October, 1961, and between that date and the final launch of the series on 30 July, 1965, NASA scored an unprecedented 100 per cent success rate (10 launchings out of 10). The sixth flight, on 28 May, 1964, carried the first model of the Apollo spacecraft into orbit.

Saturn I had a first stage made up of eight engines which burned kerosene and liquid oxygen and generated a total thrust of about 680 tonnes. The second stage contained six liquid hydrogen/liquid oxygen motors and generated 40 tonnes of thrust. The Saturn I was capable of placing 10 tonnes of genuine payload in orbit, although the seventh firing placed a total of 17 tonnes into orbit, inclusive of the spent upper stage.

A more powerful version, Saturn IB was developed to test the individual components of the Apollo spacecraft in Earth orbit. The eight first-stage engines were uprated to provide 730 tonnes of thrust, while the second stage – designated S–IV B – comprised a single liquid hydrogen/liquid oxygen motor generating a thrust of more than 90 tonnes. This increased the payload to orbit capability to about 18 tonnes. After four test flights with unmanned Apollo spacecraft, the first manned flight – Apollo 7 – began on 11 October, 1968. Had it not been for the tragic and disastrous launchpad fire which killed astronauts Grissom, White, and Chaffee, that first orbital flight almost certainly would have taken place quite early in 1967.

The vehicle needed for the actual lunar mission, the giant Saturn V, was the largest and most powerful launch vehicle that the world has yet seen. The first stage was powered by five new kerosene/liquid oxygen engines and itself stood 46 metres tall by 10 metres wide, and carried over 2000 tonnes of propellant. Each engine developed 680 tonnes of thrust and consumed over 2.5 tonnes of propellant per second. The total thrust developed was 3,400 tonnes – nearly a hundred times as much as that of the Redstone rocket which launched the first American astronauts on their sub-orbital flights. The second stage, powered by five liquid hydrogen/liquid oxygen engines, developed 520 tonnes of thrust, while the third stage was an S–IV B with a single engine producing 100 tonnes of thrust.

With a complete Apollo spacecraft and launch escape tower on top, the Saturn V–Apollo complex stood 111 metres tall and weighed about 2,900 tonnes (depending upon the precise configuration).

The first flight of the monster rocket took place successfully on 9 November, 1967 when an unmanned Apollo CSM (see below) was orbited and retrieved.

The development of the Apollo spacecraft was, if anything, a more complex task than the development of the Saturn V launch vehicle. The spacecraft consisted of three principal components: the Command Module (CM), the Service Module (SM) and the Lunar Module (LM). The configuration of Command module joined to Service Module was denoted by CSM. The Command Module housed the three-man crew on the outward and return journeys to and from the Moon, and was the only part of the complex to return to the Earth's surface. Conical in shape, with an ablative heat shield on its base, it measured 3.5 metres tall by a maximum of 3.9 metres wide; with three crew couches and the control systems aboard, it weighed about 5.5 tonnes.

The Service Module contained support systems for the CM, propellant, and the all-important rocket motor necessary to insert the spacecraft into lunar orbit and to blast it out of lunar orbit for the return journey to Earth. Its single engine provided a thrust of 9.3 tonnes. The SM on later Apollo missions weighed 25 tonnes at launch.

The Lunar Module was a two-stage vehicle weighing about 15 tonnes and was made up of a landing stage and an ascent stage. A spidery-looking structure, it measured 7 metres tall by 9.5 metres wide (measured diagonally across the span of its four landing legs).

The descent engine brought the LM from lunar orbit to a controlled landing on the surface, and the descent stage as a whole acted as a launch platform for the ascent stage which returned the two-man landing crew to orbit to rejoin their colleague, the CM pilot. Ungainly in appearance, and completely lacking in any streamlining since it would never have to operate in an atmosphere, the lunar vehicle bore some striking resemblances, in concept, to the hypothetical Moon ship which had resulted from a far-sighted design study first published in 1939 by the British Interplanetary Society.

The sequence of operations in a typical lunar mission were as follows. After launching, the third stage of the Saturn V would place the CSM and LM into a parking orbit round the Earth then, after one revolution, if all was well, would re-ignite to boost the spacecraft out of orbit and place it *en route* for the Moon. During the flight up through the atmosphere the LM was stowed in a protective canopy between the third stage and the CSM and, once the spacecraft was injected into its flightpath towards the Moon, it was necessary for the CSM to separate from the third stage, turn through 180 degrees and dock the conical end of the CM to the LM. The docked CSM then backed off and extracted the LM from the upper stage, rather like a dentist extracting a tooth.

Three days into the mission the spacecraft would swing behind the Moon and the CM engine would be fired for a six-minute 'burn' to insert the craft into an elliptical orbit round the Moon; this operation had to be carried out while the spacecraft was out of contact with the mission controllers – radio communications being blotted out by the globe of the

Moon. The next stage was for two of the crew to transfer through the interconnecting hatch into the LM, leaving the CM pilot on his own. After separating from the CM, the LM engine would be fired to reduce the LM's orbital velocity and allow it to begin to descend towards the lunar surface. After coasting down for about an hour, the LM descent engine would be restarted to slow the descent and make a gentle touchdown about 12 minutes later.

After completing the surface EVA the crew would re-enter the LM, fire the ascent engine, rendezvous and dock with the orbiting CSM. After completing the transfer of crew and lunar samples to the CM, the LM would be jettisoned to crash into the lunar surface. The next crucial burn of the SM motor would again take place on the far side of the Moon to break the spacecraft free of the lunar gravitational influence and place it into the return trajectory heading for the Earth.

After a flight lasting some 60 hours, and about 5 minutes before re-entry, the SM would be jettisoned to burn up in the atmosphere. The CM itself, approaching at nearly 40,000 km/hour had to re-enter the atmosphere at just the right angle – too steep an approach and the spacecraft would burn up, too shallow and it would skip off the atmosphere and fly off into space, never to return. The aerodynamic properties of the CM ensured a relatively gentle deceleration to the stage at which the parachutes would open; touchdown would occur in the Pacific Ocean at a point about 2000 kilometres down range from the point at which the atmosphere was re-entered.

It was a complicated sequence of events.

Apollo 7, launched by a Saturn IB on 11 October, 1968, and crewed by Walter Schirra, Donn Eisele and Walter Cunningham made 163 orbits over a period of 11 days to test out the CSM in Earth orbit. The mission was a complete success, apart from the fact that all three crew members succumbed to heavy colds while aboard! Spurred on by concern that the Soviet Union might be planning a manned lunar fly-by or orbital mission before the first Apollo lunar landing, NASA took the bold decision to go for a manned lunar orbital mission with Apollo 8. The possibility of doing this if Apollo 7 proved to be completely successful had been mooted in August, 1968, and the decision to go ahead was finally taken early in November. Prior to this decision, Apollo 8 had been planned as a manned test flight of the Apollo–Saturn V combination, to Earth orbit only. It was a gamble – not only would Apollo 8 be the first manned spacecraft to leave Earth orbit, but it would also involve the first-ever manned flight launched by the giant Saturn V vehicle.

American anxieties that the Russians might yet pip them at the post were heightened by the two Zond missions in the autumn. Zond 5, launched on 15 September, looped round the Moon and splashed down in the Indian Ocean; it was a heavy spacecraft, essentially similar to the manned spacecraft of the Soyuz series. Zond 6, 'crewed' by a variety of creatures, including miniature turtles, made a similar flight in November, and made an aerodynamic re-entry which allowed the craft to enter the atmosphere, then skip upwards to cool the heat shield before following a long, gently curving descent path to a touch-down on land. This was a sophisticated re-entry which reduced the g-forces of deceleration to a level which would be acceptable to a human crew. Speculation was rife that the Russians would try for a manned lunar fly-by in December – but no such mission occurred.

On 21 December, 1968, Colonel Frank Borman, Captain James Lovell and Major William Anders settled into the Command Module on top of the Saturn V at launch complex 39 of the Kennedy Space Center, Cape Canaveral, and the world watched on satellite-relayed television as the count-down proceeded towards zero. At 07.51 Eastern Standard Time the five engines of the first stage ignited and with an ear-splitting roar the giant 2,700-tonne vehicle surged upwards to begin an historic mission. Apollo 8 was successfully injected into lunar orbit on Christmas Eve. Borman, Anders, and Lovell completed 10 lunar orbits, taking valuable colour photographs, relaying spectacular television pictures, and describing for the Earth-bound millions the awesome scene which lay below. On Christmas morning, Borman touched the hearts and minds of most of us with his reading of the 'In the beginning . . .' passage from Genesis.

Shortly afterwards the SM motor was fired again to place the astronauts on the homeward leg. Splash-down took place

6 days 3 hours after lift-off, within 5 kilometres of the waiting carrier U.S.S. *Yorktown*. Man's first circumnavigation of another world was complete.

The key element in the lunar landing – the Lunar Module – had not yet made a spaceflight. Apollos 9 and 10 tested the vehicle – the former in Earth orbit and the latter in lunar orbit. On the Apollo 10 mission, which lifted off on 18 May, 1969, Thomas Stafford, Eugene Cernan, and John Young took the complete LM/CSM combination into lunar orbit, and Stafford and Cernan took the LM to within 15 kilometres of the surface in a full-dress rehearsal for a lunar landing. Many of us watching the progress of the mission wondered if the 'Go' might be given for a landing, but NASA was taking no chances, and Apollo 10 returned as planned, to splash-down. All the elements of Apollo had now been tried and tested, and even as the Apollo 10 crew was being picked up, the next Saturn V vehicle was on its way from the giant, 160-metre-tall Vehicle Assembly Building where vehicles are put together and mated to their payloads, to launchpad 39A, ready for the Apollo 11 mission, the first scheduled attempt at a lunar landing.

At 07.00, on 16 July, 1969 mission commander Neil Armstrong, Command Module pilot Michael Collins, and Lunar Module pilot Edwin 'Buzz' Aldrin took up their positions in the Command Module, 'Columbia'. The minutes ticked away relentlessly as the multitude of checks continued methodically. At T–200 seconds the countdown became automatic. 9 seconds before lift-off the first of the five motors burst into life, followed soon after by the others. Flame and smoke spewed from the base of the launchpad as, held by restraining arms, the giant Saturn V stood motionless – roaring like a demented dinosaur – for the few seconds which it required for full thrust to be built up. At T = 0 the hold-down arms were released and, slowly at first, but ever more rapidly, the 2,900-tonne vehicle rode a column of flame into the sky. The familiar, but welcome, message came over the air, 'Lift-off! We have lift-off!'

Apollo 11 was on its way.

On 19 July Apollo 11 entered lunar orbit, and the following morning Armstrong and Aldrin entered the Lunar Module to

Plate 14. Lift-off of Apollo 11 from Cape Kennedy on 16 July, 1969 at the start of the historic lunar-landing mission. (NASA/Woodmansterne).

begin their descent. As they approached the target area in the south-western corner of the Mare Tranquillitatis (Sea of Tranquillity) it became clear that the designated landing site contained a rock-strewn crater. Armstrong took over manual control, his heartbeat rising to twice its normal rate. Guided by Aldrin, who was giving altitude and velocity readings, Armstrong flew the flimsy craft to touch-down at 16.18 EDT; As the Lunar Module 'Eagle' touched down on the lunar soil, there remained only twenty seconds' supply of fuel for the descent engine. Back at mission control, Houston, after flashing across nearly 400,000 kilometres of space in 1.3 seconds, a welcome message was received from Armstrong: 'Houston, Tranquillity Base here. The Eagle has landed.' Man had reached the Moon, and the news was received with relief and elation by Mission Control and by the largest global television audience in history.

6 hours 21 minutes later, Armstrong squeezed out of the LM hatch and moved slowly down the nine steps of the ladder until, at 22.56 EDT his left foot touched the lunar soil and a man had arrived for the first time on the surface of another world. A television camera on board the Lunar Module relayed the scene to the watching world. Although the picture quality of that first 'small step' was dreadful by comparison with later Apollo missions, those of us in Britain who sat up through the night to watch the event (it occurred at 03.56 British Summer Time) knew full well that we had been privileged to witness a unique event unparalleled in human history.

Some 20 minutes later Armstrong was joined by Aldrin on the dusty surface of the Sea of Tranquillity, and both men bent themselves to the various scientific tasks they had been assigned: the collection of rock and soil samples; the deployment of a strip of aluminium foil to detect particles of the solar wind, of a seismometer to detect moonquakes and meteorite impacts, and of a laser reflector to be used to reflect laser beams back to Earth to make the most precise possible measurements of distance.

The American flag (held up by a wire insert to prevent it from sagging in the absence of air and wind) was planted, and Armstrong and Aldrin received a congratulatory telephone call from President Richard M. Nixon, and the rest of us watched in awe.

By 01.09 EDT on 21 July, the EVA was over, Armstrong and Aldrin having returned to the Lunar Module. At 13.54 the

Plate 15. Edwin Aldrin descends the steps of the Apollo 11 Lunar Module 'Eagle' on 20 July, 1969, to become the second man to set foot on the Moon. (NASA)

ascent engine was fired and 'Eagle' surged upwards from the descent stage to re-dock with Columbia at 17.35. Back on the Moon, all that remained were assorted artifacts and the spider base of the Lunar Module, a plaque fixed to one of its legs bearing this message:

"Here men from the planet Earth
first set foot on the Moon
July 1969 A.D.
We came in peace for all mankind"

The return journey went without incident and Columbia splashed down in the Pacific Ocean at 12.51 EDT (17.51 BST) on 24 July to bring to a conclusion an epoch-making voyage of exploration. After a helicopter ride to the recovery ship U.S.S. *Hornet*, the crew entered a quarantine chamber in which they were to remain until 21 days after leaving the lunar surface. This was a precaution against the extremely remote possibility that they had brought some infectious organism from the Moon – an undignified but sensible precaution (although it must be said that since they had already been in contact with the helicopter crew and the welcoming party on the carrier – the effectiveness of the quarantine procedure was limited).

So ended Man's triumphant first visit to the Moon.

Five more Apollo missions, number 12, 14, 15, 16, and 17 landed on the lunar surface and returned safely, each extending the scientific scope of their activities, the distance traversed on the lunar surface, and the quantity of lunar samples returned. Apollo 12 lifted off with high drama on 14 November, 1969, the launch vehicle being struck by lightning shortly after leaving the launchpad, but, miraculously, no damage was done and the mission proceeded as planned. Remarkably, the lunar module 'Intrepid' was able to touch down just 180 metres from the unmanned Surveyor 3 spacecraft which had landed two and a half years previously and astronauts Charles Conrad and Alan Bean were able to walk across, make an inspection, and retrieve parts of the spacecraft for analysis back on Earth.

High drama came with Apollo 13 which lifted off from Cape Canaveral on 11 April, 1970 with mission commander James Lovell, together with Fred Haise and Jack Swigert – a last-minute substitute for Thomas Mattingley who had been exposed to measles prior to the launch date. Just under 56 hours into the mission, when the spacecraft was some 330,000 kilometres from Earth, Swigert suddenly called to Mission Control, 'Hey, we've got a problem here' – a masterpiece of understatement as it turned out. An explosion in the Service Module oxygen tank had severely damaged the module and knocked out the power supplies necessary to sustain the Command Module. All thoughts of a lunar landing were abandoned – and all efforts were directed to the urgent and dramatic task of saving the astronauts, and bringing them back to Earth.

Fortunately, the Lunar Module systems were independent, and so the LM 'Aquarius' was able to act as a lifeboat, using its life-support and power supplies to sustain conditions for the astronauts at minimum tolerable levels. The descent engine of Aquarius had to make three crucial burns to get the spacecraft into a path which would loop round the Moon and return to Earth, and on each crucial occasion the LM engine performed flawlessly. As they neared the Earth, the crew jettisoned the SM and, just prior to re-entry, the LM – which had saved the crew – had to be cast adrift to burn up in the atmosphere. The Command Module 'Odyssey' made a faultless re-entry and splashed down within sight of the waiting flotilla of ships.

What could have been a disaster of catastrophic proportions, not only for the crew but for the whole Apollo programme, had been turned into a triumph by coolness, resourcefulness and ingenuity, and by the flexibility of the LM itself. One very sombre thought must remain. The explosion which incapacitated the SM occurred on the outward flight; had it occurred after the LM had made its descent, nothing could have saved the crew. The LM would not have been available as a lifeboat, and the crew would have been doomed to perish in the lonely recesses of space.

Apollos 14 to 17 all made successful landings, exploring progressively rougher terrain, on each occasion increasing the duration of EVA and the quantity of rock and soil samples returned. On Apollo 11, EVA time amounted to 2.2 hours and the weight of samples returned was about 20 kilograms. On the final mission, crewed by Eugene Cernan, Ronald Evans (CM pilot) and Harrison Schmitt – the first trained geologist to make a lunar landing – EVA time totalled 22 hours and 112 kilograms of samples were brought back.

Apollos 15 to 17 carried with them the Lunar Roving Vehicle (LRV), a four-wheel, battery-powered vehicle, and a miracle of lightweight engineering, which allowed the astronauts greatly to increase the range of surface exploration. Driven with 'élan' it reached a maximum speed of 17 km/hour and, on Apollo 17 covered a total distance of 29 kilometres (although at no time did the Lunar Rover travel further from the LM than the astronauts were capable of walking – about 9 kilometres – should the LRV break down).

As a result of Project Apollo we now have a much deeper and far more comprehensive understanding of the Moon – its surface, age, and internal structure. Lunar rocks have proved to be similar to those found on Earth, the mare regions containing basalts, and the uplands anorthosites – similar in nature to the rocks found in some mountainous regions of the Earth. The principal difference between lunar rocks and terrestrial ones being the higher fraction of 'refractory' elements (i.e. elements with high melting points) such as aluminium and titanium in the former. Analysis of moonquakes – deep seated but relatively modest internal events – and meteorite impacts (natural and man-made), together with heat-flow experiments, have revealed an internal structure very different from what had been expected; a central region which is hot (about 1300 K and probably partially molten) overlain by a rigid layer of rock about a thousand kilometres thick on top of which, in turn, is the crust – thicker on the far side than on the Earth-facing hemisphere.

The oldest surface rocks found by the astronauts are about a billion years older than anything yet found on Earth – between 4.4 and 4.5 billion years old, and date from a period very close indeed to the formation of the Solar System. That the majority of craters were formed by impacts during a period of heavy bombardment in the first billion years of the Moon's existence now seems well established. The mare, almost certainly were formed by giant impacts which scooped out basins tens of kilometres deep; as the lunar interior heated up, between 3.8 and 3.1 billion years ago, lava welled up to fill in the basins, giving the dark mare regions which we see today. During the past 3 billion years no major changes have occurred, apart from a small number of massive impacts giving rise to craters like Copernicus, and some minor residual internal activity.

Apollo 17 brought the series to a close. Although the hardware existed to send three further missions, it was decreed that no further missions would fly. One of the three remaining

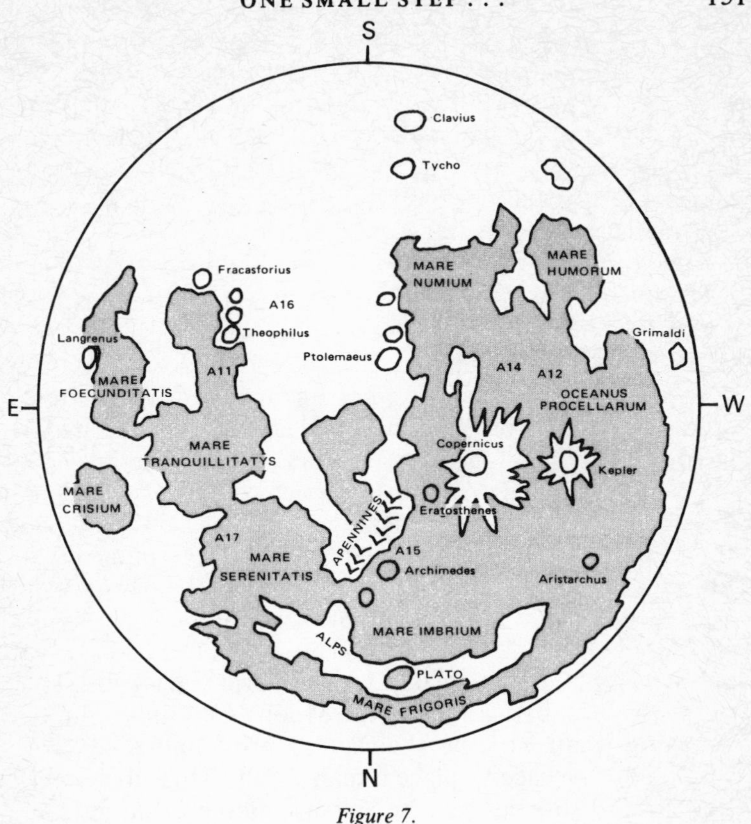

Figure 7.

Landing sites of Apollo spacecraft. Apollo 11 is indicated by *A11*, Apollo 12 by *A12*, etc.

Saturn Vs was used to launch Skylab (see Chapter 10) in 1973, but the other two have become, in effect, museum pieces – wistful reminders of the heady days of Man's dramatic quest for the Moon, and testifying to the fact that by the early 1970s – at least in the eyes of the American administration – the gloss had worn off the manned space programme, and the visionary zeal which had driven men to try for the Moon, had lost its momentum.

On the day that Apollo 11, landed, London bookmakers William Hill were offering odds of 100:1 against a manned Mars landing before 20 July, 1976, and 'evens' on such a landing before 20 July, 1979 (the tenth anniversary of the

Moon landing)! The renowned radio astronomer Sir Bernard Lovell, writing in 1969, remarked 'technically a human landing on Mars in 1980-85 is probably more realistic than the 10-year scale for moon landing was in 1961', and Wernher von Braun commented that 'If our current manned spaceflight continued to evolve . . .' – manned planetary exploration would be feasible in the 1980s.

Alas, such predictions were premature. Given the impetus, and the enormous financial backing necessary there is little doubt that it would have been within the scope of Man's ingenuity to achieve that aim. But the economic and political climate had turned against extravagant manned space programmes. Even before Apollo 11 had landed, NASA's budget had begun to suffer progressive cuts. When Vice-President Spiro Agnew – while Apollo 11 was en route to the Moon – declared his view that the next major target should be to place a man on Mars before the year 2000, he was immediately subjected to harsh criticism by a number of senators who argued, reasonably enough, that the needs of the Earth should be placed before a Mars landing. The magic, for the moment, had gone, and the harsh realities of the world had taken over.

Finally, was there a race to the Moon? Certainly from the American point of view the Apollo programme was seen very much as a race to place a man on the Moon before the Russians did so – but was the Soviet Union ever in the 'race' at all? It has become fashionable in the West to consider that the Soviets were never in a race as such and were more concerned with long-term objectives in space, in particular, with the establishment of a large, orbiting space station as a pre-requisite to manned flights to the Moon and planets. The steady progression of Soviet Earth-orbital activity in the past decade may seem to support that view. It is also clear that to achieve a manned landing, the Soviet Union would have required a launch vehicle far more powerful than any in their possession in the nineteen-sixties (the 'G–1' perhaps?). Despite this, however, Soviet cosmonauts, up to 1967 or so, often remarked that when the Americans reached the Moon, Russians would be there to greet them.

The flights of Zonds 5 and 6, already alluded to, point strongly to the possibility that the Soviet Union was planning a manned fly-by and return mission which was pre-empted by the successful flight of Apollo 8, itself a mission staged at short notice to attempt to ensure that the Soviet Union would not be first to lunar orbit. James Oberg, author of *Red Star in Orbit* is fully convinced that the Soviet Union was in the race right up to the last minute, and suggests a plausible scenario whereby a manned craft launched by the Soyuz launch vehicle (A–2) could have been mated in orbit to a propulsion stage launched by the more powerful Proton (D–series) rocket, and then gone on to achieve a full lunar orbital mission. Whether or not such plans existed and were thwarted by technical troubles, we may never know.

Whether or not there was a race to the Moon, the fact which remains is that in July 1969, the goal was attained, and men set foot on the Moon. Regardless of the politics involved, 20 July, 1969 will stand for ever as the date when one man – Neil Armstrong – took that 'one small step' which irrevocably altered mankind's status in the universe and opened the door to wider adventures of exploration.

10

Space Stations and Space Meetings

Tsiolkovsky, far-sighted prophet of the Space Age, was the first to write convincingly of the possibility of housing entire communities in cities in the sky, and Hermann Oberth was another who foresaw the potential applications of a space station as a platform for observing the Earth, and the Universe, and as a base for interplanetary flight. An outline design for a wheel-shaped space station, spinning on its axis to generate a sensation of artificial gravity at its perimeter, was produced in 1928 by the Austrian, Captain Potocnik.

Many others have considered the possibility since that time, Wernher von Braun being a particularly enthusiastic advocate who considered the design, mode of construction, and methods of servicing such a station, and advocating an 80-metre diameter wheel as the basic structure. Writing in 1952, he remarked, 'Development of the space station is as inevitable as the rising of the Sun . . . ' and went on to suggest that a manned station of this type might be constructed by about 1963. In the event, practical progress toward the realization of this goal has advanced at a more modest pace, but there can be no doubt of the importance of establishing large space stations if mankind is ever to become totally at home in, and exploit fully the potential of his space environment. The Soviet Union has consistently declared the establishment of such stations to be a major goal of its space programme.

The tragic death of Vladimir Komarov occurred at the end of the first manned test flight of the Soyuz spacecraft, a spacecraft which has now been the backbone of the Soviet manned programme for well over a decade. Soyuz 2 was flown as an unmanned test flight on 25 October, 1968, eighteen months after the Soyuz 1 disaster, and the following day Giorgi Beregovoy was launched aboard Soyuz 3, and made rendezvous with Soyuz 2, first on automatic control, and again under manual control. Touch-down occurred right on target in Kazakhstan – all the spacecraft systems performing successfully this time.

The Soyuz spacecraft is much more sophisticated than its Vostok and Voskhod predecessors, and consists of an orbital module (OM), a descent vehicle (DV) – the only part to return to Earth – and the Instrument-Assembly Module (IAM) which includes the orbital propulsion systems. The OM measures about 2.2 metres in diameter and is used for experimental work and as a rest and sleeping area for the crew while in orbit. Mounted at the front of the spacecraft it contains the docking mechanisms and transfer hatch necessary to transfer crews to docked spacecraft or space stations. Separated from the re-entry module by an internal hatch, the OM can be used as an airlock in crew transfers or for EVA. The descent vehicle provides accommodation for up to three cosmonauts during launch and re-entry operations. Coated with heat-resistant material, and with an ablative heat shield on its base, it has an aerodynamic shape which gives 'lift' and allows much gentler deceleration on re-entry than was the case with the earlier Soviet manned spacecraft.

The overall length of the complete spacecraft, including the propulsion systems of the IAM is 7.5 metres and the weight, 6.8 tonnes. Two wings carry solar panels to produce electrical power, and give the spacecraft an appearance not unlike a flying ant! The total habitable volume while in orbit is 10 cubic metres.

Soyuz spacecraft of this basic pattern have been used from 1967 up to 1981 when it was stated by Soviet officials that Soyuz 40 marked the last launch of the old-style craft. A more advanced version – the T–series – has made four flights up to the time of writing (early 1982), the most recent being Soyuz

T–4, launched on 12 March, 1981 . . . The general dimensions of this newer version are the same as for the older Soyuz; the main differences are thought to lie in lighter and more compact onboard systems.

The first significant success of the Soyuz programme came in January 1969 when Soyuz 4, piloted by Vladimir Shatalov, docked with its sister ship Soyuz 5 with three cosmonauts aboard – Boris Volynov, Yevgeny Khrunov, and Alexei Yeliseyev. The most novel aspect of the mission was the crew transfer which took place. Khrunov and Yeliseyev stepped out into space through a hatch in the side of the orbital module of Soyuz 5, and made their way along the outside of the linked spacecraft to enter Soyuz 4 after a 37-minute EVA. After demonstrating what was in effect a technique for space rescue, Soyuz 4 and 5 separated, the former landing with three crew aboard and the latter with one as a result of the 'switch'.

In October of that year Soyuz spacecraft 6, 7, and 8 were all in orbit at the same time – the first time that three spacecraft had been orbiting simultaneously – and carried out various scientific and technical tasks, including experiments in welding in zero gravity – an activity regarded as essential for the assembly of large space structures. In June 1970, Soyuz 9 regained for the Soviet Union the space endurance record, which had been held since December 1965 by Gemini 7, by remaining in orbit for nearly 18 days.

The next major step came on 19 April, 1971, when, just ten years and one week after Gagarin's flight, a Proton launch vehicle lifted a 19-tonne 'Salyut' space station into a low Earth orbit at an altitude of about 200 kilometres. It was 16 metres long, with a maximum diameter of 4.15 metres and a habitable volume of about 100 cubic metres, and was the largest payload yet placed in orbit by the Soviet Union, and the world's first genuine orbiting laboratory, or space station.

Three days later, Soyuz 10 docked with the Salyut, but the three-man crew did not, as was widely expected, transfer to the space station. That task fell to the crew of Soyuz 11, which docked with Salyut on 7 June 1971. The interconnecting hatches were opened and the crew transferred to the more palatial surroundings of the space station. A long-standing

Plate 16. Mock-up of a docked Soyuz spacecraft (top) and Salyut space station (bottom). (Novosti Press Agency)

dream, of ferrying men to an orbiting space station – albeit on a fairly modest scale – had at last been achieved. The first two days were spent commissioning the space station and thereafter the cosmonauts settled into a routine of observation, tests, and experiments. Daily television broadcasts to the U.S.S.R. allowed millions of Soviet citizens to see the cosmonauts in the orbiting laboratory which was to be their home for 23 days.

On 29 June they returned to their spacecraft, and at 21.28 Moscow time, Soyuz 11 separated from Salyut. About four hours later the retro-motors fired to start the descent into the atmosphere, and at that point communication was lost. All the automatic landing systems functioned perfectly to land the capsule gently on Soviet soil. Then came the announcement which made headline news all over the world – the three cosmonauts were dead, still strapped in their seats and without obvious signs of a struggle. The mysterious deaths of the three cheerful astronauts whose space activities had been watched by millions produced a reaction of shock and sadness all over the globe.

Immediately speculation began as to the cause of the disaster. One theory which came at once to the fore was that prolonged exposure to weightlessness – five days longer than any previous spaceflight – had weakened the cosmonauts' bodies to such an extent that the forces experienced in re-entry had killed them. The London *Evening News* of 30 June carried the headline 'Space Death Plunge' and quoted an 'exclusive' report from Moscow that '. . . scientists here are inclined to believe that "failure" of their bodies may have been to blame, having been brought on by an unprecedented 23 days in weightlessness in orbit.' Inevitable questions were asked . . . 'Were they in space too long?,' 'Is it worth it?'.

The explanation was soon revealed, however. A faulty valve had allowed air to escape rapidly from the descent module after it had separated from the rest of the Soyuz craft. There was nothing the cosmonauts could do. They were dead within minutes.

On October 11 Salyut re-entered the atmosphere and broke up.

Salyut 2 was placed into orbit on 3 April, 1973, but after orbital manœuvres carried out eleven days later, the space station tumbled out of control and by the end of May had broken up. The Salyut programme seemed to be ill-fated.

In the United States various schemes for manned space stations had been discussed from time to time, and in the early nineteen-sixties several design studies had proposed using the spent upper stage of a launch vehicle to be fitted out – after its fuel had been consumed – as a manned laboratory; this was the so-called 'wet' laboratory concept. Two basic contenders emerged, one military, and one civilian.

The military project was the Manned Orbiting Laboratory (MOL), proposed by the U.S. Air Force, and given official sanction in 1965. The idea was to use the Titan II vehicle as the basis, and to fly the laboratory in a low polar orbit for reconaissance purposes; as the Earth rotated beneath its orbit, all parts of the globe would be accessible to the crew. MOL was cancelled by the Nixon administration in 1969.

The civilian project, originating with NASA, was to build a space station based on Apollo–Saturn hardware. In the first 'wet' concept the S–IVB upper stage of a Saturn IB launcher would be fitted out in orbit by later flights. In August 1969, after the successful Apollo 11 lunar landing, the decision was taken to divert one of the Saturn V vehicles to the orbital workshop and that the 'wet' concept could be abandoned. The complete laboratory (basically an S–IVB fitted out on the ground), with its crew compartment, docking adaptors, telescope mount, and all other facilities could be assembled 'dry' on Earth and placed in orbit in one piece. Thus 'Skylab' was born.

Skylab was the largest piece of hardware ever to be put into space. Including a docked Apollo CSM, it had an overall length of 36 metres and a weight of just over 90 tonnes; the standard Apollo CSM was used as the ferry craft. Skylab consisted of the following principal elements: the orbital workshop which provided crew's quarters and experimental areas and had a habitable volume of 295 cubic metres – about three times more than the Soviet Salyut; the multiple docking adaptor which provided a capability (not used in practice) for docking two Apollo spacecraft simultaneously, an airlock module connecting the docking adaptor to the workshop, and the Apollo Telescope Mount, carrying telescopes and other instruments for studying the Sun. The main body of the workshop was designed to carry two wing-like solar panels, exposing 219 square metres of surface to the sunlight for the production of electrical power. The telescope mount carried its own array of solar panels. The solar panels were folded during launching, and the whole of Skylab housed inside a protective shroud for its journey up through the atmosphere.

Precisely on time, on 14 May, 1973, Skylab lifted off from Cape Canaveral. One minute into the flight, trouble began. A large chunk of the micrometeorite shield broke loose, tearing loose one of the workshop's solar panels (this panel eventually ripped off altogether) and jamming the other so that once Skylab was in orbit, it refused to deploy on command from the Earth. To make matters worse, the micrometeorite shield was also intended as a heat shield to keep direct sunlight off the laboratory and maintain an equable temperature inside. Skylab entered its planned orbit at an altitude of 435 kilometres, but it was seriously crippled and soon was overheating severely. To counteract this, Skylab was tilted at an angle of 45° to the Sun, but this lowered the efficiency of the remaining solar panels to such an extent that barely enough power was being generated to keep essential systems going.

The situation looked bleak. Was the billion-dollar piece of hardware – the first (and so far the only) American space station – destined to become a write off before the first crew had even arrived?

The first manned mission to Skylab had been scheduled for the following day, but was put off for 10 days, and in that time NASA engineers and the astronauts worked flat out to devise the means to save Skylab. When the first crew of Charles Conrad, Joseph Kerwin, and Paul Weitz lifted off on 25 May, their primary task was to act as repair men.

After a visual inspection carried out by flying around Skylab in their CSM, and after a brief pause for a meal, the crew first attempted to free the jammed solar panel by means of a pole operated from the CM hatch. When this failed they re-docked, with considerable difficulty (in fact it was necessary to go out of the spacecraft to partially dismantle and re-assemble the docking mechanism before a successful docking was achieved) and made their first excursion into the laboratory. This had to be done with caution, wearing gas masks in case toxic fumes had been released from overheated insulation material; although the temperature inside was about 55°C (130°F), all seemed to be well.

The next task, successfully accomplished, was to deploy a 'parasol' thermal shield outside the workshop; with this in position, the temperature in the workshop immediately began to drop. Then came the most difficult task. With the aid of specially developed tools – produced in those 10 frantic days – and using techniques practised in a tank of water containing a full-size mock-up of the workshop, and which simulated to a certain extent the problems of working in weightless conditions, they attempted to free the jammed solar panel. Working in space with a cable cutter attached to a 25-foot aluminium pole, Conrad and Kerwin struggled at their difficult, dangerous, and exhausting task, while back on Earth, their efforts were mimicked in the simulator and advice (not all of it kindly received!) relayed to space. Several times, as the cutter severed straps and brackets, the sudden movements sent the astronauts tumbling to the limit of their tether lines, but in the end they succeeded, and the wing deployed fully, to double the power supply to the laboratory, and ensure the viability of the Skylab mission.

In the most emphatic way possible, Conrad, Kerwin, and Weitz had demonstrated Man's ability to work in space, to carry out repairs, and to rescue from the brink of disaster a major technological project. Those who doubted the value of man versus machine in space were most effectively silenced.

The Skylab mission proved to be highly successful thereafter. The laboratory itself provided standards of comfort previously unheard of in space, with all 'mod cons' including a zero-gravity 'shower', exercise machines, and a standard of catering which was far superior to anything seen in earlier missions – turkey and gravy even! With the enormous volume, there was ample room for weightless gymnastics, and the view from the window was agreed by all to be utterly breathtaking.

In all, three crews, each of three astronauts, spent a total of 513 man-days in space carrying out a wide range of observations and experiments. The areas which received most attention were solar astronomy (31 per cent of the total time allocation), Earth observations (19 per cent), astrophysics (9 per cent), and life sciences (27 per cent). Work was also carried out in engineering technology and materials science. Student experiments, selected by competition, were carried aboard, one of the most novel being a test to find out if a spider could still spin its web in zero gravity; the spider,

Plate 17. The Skylab space station photographed in June 1973 by the crew of the first manned mission prior to their return to Earth after successfully carrying out vital repairs, including the deployment of the one remaining 'wing' of solar panels on the main body of the orbital workshop, and the erection of the 'sunshade'. (NASA)

'Arabella' was disorientated at first but soon readjusted and spun away happily!

Crystals of a size and perfection unattainable on Earth were grown, and experiments with alloys performed, such experiments indicating the potentiality of space as a manufacturing facility – a line which has consistently been pursued in the Salyut missions. But without doubt the most fascinating scientific results came from the solar studies (carried out by viewing the Sun in ultraviolet and x-ray-radiation which is inaccessible to ground-based observers) which played a major role in developing the modern view of the Sun, touched upon briefly in Chapter 6.

Considerable information, too, was gained about the adaptability of the human being to weightlessness, during progressively longer stays of 28, 59, and 84 days for the three Skylab missions, each of which in turn established a new space endurance record. The human body was shown to adapt well to its new environment, any physiological changes levelling off after a time and returning to normal after the astronauts had returned to Earth. One slightly disturbing aspect was a long-term loss of bone calcium which may yet lead to problems for astronauts returning from extremely long stays in space, but that is a question which remains to be resolved.

Skylab itself re-entered the atmosphere and broke up on 11 July, 1979, scattering a hail of fragments across the Indian Ocean and parts of Australia. This event aroused enormous world-wide publicity and concern at the time, although, in fact, no damage was done to persons or property, and the risk of such damage occurring was minimal. No such furore surrounded the re-entry and break-up of other heavy payloads, such as the Soviet Salyuts (admittedly much less massive than Skylab), but Skylab really hit the headlines in a big way. Ironically, in view of the fact that one of its principal aims was to study the Sun, solar activity was largely responsible for bringing about the demise of Skylab much earlier than NASA planners had orginally expected. As the Sun's cycle of activity moved towards its peak in the late nineteen-seventies, one of its effects was to increase the height of the atmosphere, and so increase the atmospheric drag on the space station.

While the Skylab mission was under way, discussions were continuing between Soviet and American space officials about a proposed joint manned flight – The Apollo–Soyuz Test Project (ASTP) – in which an orbiting Soyuz would act as a docking target for an American Apollo CSM. An agreement to undertake the development of mutually compatible docking mechanisms to facilitate the possibility of space rescue and to permit joint scientific missions, with the aim initially of a joint Apollo–Soyuz flight in 1975 was signed on

24 May, 1972 by Richard M. Nixon for the United States and Aleksei N. Kosygin for the Soviet Union. The optimists among us were delighted, regarding this as the first step in opening up the possibility of making space exploration a fully international activity – and parallels were drawn with the exploration of Antarctica. Others were more sceptical, and in the event the flight took place very much as a 'one-off', a political gesture which, so far, has not led either to further joint missions or to any plans for such missions. Nevertheless, a door has been opened, if only by a crack, and joint missions have been shown to be possible, at least.

Thus it was that at 20.19 GMT on 17 July, 1975, Astronaut Thomas P. Stafford, and cosmonaut Alexei A. Leonov shook hands through the interconnecting hatch of the docked Apollo–Soyuz complex to symbolize the success of the first international link-up in space. For two days the spacecraft remained linked together. Astronauts Stafford, Vance Brand, and Donald Slayton visited the Soyuz, and Leonov and Velariy Kubasov visited the Apollo craft. At a personal level the whole mission was a great success, the astronauts and cosmonauts – who had already met on training sessions – getting on extremely well. Afterwards, the two spacecraft separated and went their own ways. So too did the Russian and American space programmes.

After ASTP, six years were to pass before Americans next went into space (see Chapter 11). The Soviet Union, however, methodically continued its programme of Soyuz and Salyut missions, building up experience, continuing its scientific and technological experiments, and developing techniques to establish a firm foothold in space. Despite early setbacks and disasters, the decade since Salyut 1 and Soyuz 11 has been one of steady, although sometimes erratic, progress.

The Salyut programme includes both military and civilian missions. It is generally reckoned that of the designated Salyuts so far launched, numbers 1, 4, and 6 were essentially civilian in nature, and 2, 3, and 5 were essentially military in nature; the 'military' Salyuts flying in lower orbits than the 'civilian' ones, presumably for reconnaissance purposes in much the same fashion as had been intended for the cancelled American MOL.

Salyut 3 was placed in orbit at a height of about 260 kilometres on 24 June, 1974, and on 3 July, Soyuz 14 lifted off, this time with two cosmonauts aboard – Pavel Popovich and Yuri Artyukhin. Their successful return to Earth after two weeks on board the Salyut laid to rest the ghastly spectre of the Soyuz 11 mission, and the way seemed clear for an ongoing series of missions. All still did not go smoothly, however, for the next mission to Salyut 3, the following month, failed to achieve a link-up.

Salyuts 4 and 5 followed in December 1974 and June 1976, still with rather mixed fortunes so far as achieving successful crew transfers was concerned, but with Salyut 6, everything came right. The space station was placed into a 350 kilometres orbit on 29 September, 1977, and at the time of writing is still in orbit. Although the first attempted link-up by Soyuz 25 failed when the docking mechanism obstinately refused to work, the Soyuz 26 mission, flown by Georgi Grechko and Yuri Romanenko and launched on 10 December, 1977, docked successfully with the additional rear docking port which was a new feature of Salyut 6. Their inspection of the forward docking mechanism showed that it, too, was in full working order, and Salyut 6 was fully operational.

Grechko and Romanenko remained aboard for 96 days, so overtaking the space endurance held by the third Skylab crew since early 1974, and during their flight notched up several notable 'firsts'. They were joined on 10 January, 1978, by the crew of Soyuz 27, so that for the first time, two manned spacecraft were docked to the same space station. The Soyuz 27 crew returned to Earth in the Soyuz 26 spacecraft, leaving the 'fresher' Soyuz 27 docked to the Salyut. Ten days later another major step towards sustaining lengthy missions on space stations was achieved when the unmanned freighter, 'Progress 1', docked to replenish stocks of food and consumables, and to re-fuel the propulsion system necessary for control of the Salyut and for the periodic burns necessary to boost it to higher altitude each time its orbit began to decay. The first non-Soviet cosmonaut to go into space was Vladimir Remek of Czechoslovakia who, with Alexsei Gubarov flew an 8-day mission to Salyut 6 with the Soyuz 28 spacecraft in March, 1974.

Since that time a regular pattern has emerged, with long-stay crews being visited regularly by others, bringing mail and fresh supplies; the short-stay crews return to Earth in the Soyuz which has already been in space for some time, leaving their freshly-fuelled craft for the long-stay crew (in this way the risk of deterioration of the Soyuz is minimized). Non-Russian cosmonauts have featured prominently among the short-stay visitors. Up to early 1982, nine cosmonauts from nine nations outside of the Soviet Union have flown Soyuz missions, the nations represented so far being Czechoslovakia, Poland, East Germany, Bulgaria, Hungary, Vietnam, Cuba, Mongolia, and Rumania. Next on the list are likely to be France and India.

Onboard activities have included astronomical observations with the onboard 'Orion' telescope system, geophysical observations, and a host of biological and physiological experiments; a multispectral camera has been used for studying the Earth at six different wavelengths for Earth resources work. There are two electrical furnaces aboard, SPLAV and KRISTALL, the former being used to heat alloys to about 1100°C and the latter to work on semi-conductor materials. On-going industrial processing is a regular and continuing feature of the Salyut programme.

On 14 May 1981, Soyuz 40, commanded by the Russian, Leonid Popov and crewed by Dumitru Prunariu of Rumania docked with Salyut 6 to join Vladimir Kovalyonok and Viktor Savinykh (the latter being the hundredth person to fly in space) who had boarded the space station in March from Soyuz T–4. Soyuz 40 returned to Earth on 22 May, followed by Soyuz T–4, four days later. Since that time, up to the time of writing, no further manned flights to Salyut 6 have taken place, but on 19 June, 1981, the heavy spacecraft, Cosmos 1267, docked with the space station and, eleven days later, fired its motor to boost the linked combination to a higher (339 × 360 km) orbit. With Cosmos 1267 – described as 'a prototype space module' – attached, the size of the space station has been virtually doubled. The true nature of Cosmos 1267 is still unknown, although a report in *Aviation Week and Space*

Technology (30 November, 1981) suggested that the new module is equipped with anti-satellite interceptor rockets.

Whatever the future role, if any, of Salyut 6 may prove to be, it has already been a most successful space station. In its first four and a half years of existence it was inhabited for a total of 676 days. Five long-stay missions were accomplished, 11 visits were made, and 12 unmanned Progress supply ships docked to replenish consumables. New space endurance records were set: 96 days, 140 days, and 175 days by the crews of Soyuz numbers 26, 29, and 32, respectively; and the overall record for the first 25 years of the Space Age (which could not be beaten before 4 October 1982) of 185 days set by Leonid Popov and Valeri Ryumin who set off aboard Soyuz 35 on 9 April, 1980, and returned aboard Soyuz 37 on 11 October of that year.

Salyut 7 was placed into orbit on 19 April, 1982, but it remains to be seen what its role will be.

New developments would appear to be in the offing. The next logical step forward, consistent with stated Soviet aims, would be the construction in orbit of a modular space station made up of two or more Salyut-sized units. Existing Soviet launch vehicles do not have the payload capacity to place in orbit a single large station like Skylab, so the modular approach is the only one open at this time; if the long awaited 'big new launcher' (G–1) does make its appearance then, of course, that situation would change.

With Skylab and the Salyut series, Man has gained his first long-term footholds in space. The assembly of large regularly manned space stations in orbit would seem to be essential to the long-term 'conquest' of space, and the development of such stations is likely to be one of the major directions for space activity in the second quarter century of the Space Age.

11

Into the Shuttle Era

A new epoch in space transportation, the era of the re-usable spacecraft, may properly be said to have arrived on 14 November, 1981, when the Space Shuttle 'Columbia' touched down at Edwards Air Force Base, California, after its *second* orbital mission. For the first time in history a spacecraft had made more than one flight into space, the first flight of 'Columbia' having taken place in April of that year. After a decade of planning, development, and constructional work, hampered by financial constraints and dogged by a multitude of problems and hold-ups, to such an extent that many were sceptical of the Shuttle ever making its way safely into space and back, 1981 was truly the Year of the Shuttle.

The need for a re-usable spacecraft had been obvious for as long as man had seriously considered the question of getting into space and establishing a permanent foothold there. Prior to the Shuttle, each launch vehicle was used once, and once only, as they were incapable of being retrieved and refurbished. This was a costly and wasteful method of getting into space; no terrestrial delivery service could hope to stay in business if they threw away the delivery van at the end of each trip!

In 1952 Wernher von Braun had discussed the construction of a giant wheel-shaped manned space station to be established in orbit at an altitude of 1730 kilometres. To construct and service this station von Braun proposed the building of a huge three-stage vehicle weighing some 7000 tonnes at launch, and each stage of this vehicle was intended to be retrieved and re-used. The first and second stages were to be parachuted into the ocean for collection by waiting surface ships, while the third stage was to be a piloted winged vehicle 23.5 metres long with a wingspan of 47.5 metres and an all-up weight (fuelled) of about 220 tonnes; its cargo capacity was to be 37 tonnes. The entire vehicle was to be assembled in a separate building and moved to the launch platform on a mobile platform, then launched in a vertical configuration to place the winged third-stage vehicle in orbit. The third stage was to re-enter the atmosphere at a shallow angle, using its wings to make a prolonged re-entry at relatively gentle rates of deceleration, gliding to a touchdown on a conventional runway. Although differing in many respects from the real Shuttle which flew three decades later, many familiar elements of the modern Space Transportation Systems are present in that early conjectural design.

Others, too, at that time, and earlier, had envisaged winged rocket vehicles which would fly to a landing like a conventional aircraft. Von Braun hinted that such a vehicle could be built by the mid-nineteen-sixties; in reality his timescale was out by a factor of two.

The origin of re-usable space vehicles has many strands in history. Rocket-assisted gliders were flown in Germany in the nineteen-twenties, and of obvious relevance were the experimental rocket-powered aircraft, particularly those developed in the United States from the late nineteen-forties through to the late nineteen-sixties. First of these was the Bell X–1 a rocket-powered craft, 9.5 metres long with a wingspan of 8.5 metres, which was carried aloft in the bomb bay of a B–29 bomber; released from its 'parent' the X–1 fired its rocket motor to accelerate itself to high speed then, when the motor cut out, it glided to a conventional landing on the runway at Edwards Air Force Base. The first powered flights took place in 1946, and in 1947 Charles E. Yeager became the first man to fly faster than sound. Further developments of the rocket plane in the nineteen-fifties raised the speed and altitude capability and, in 1959, the X–15, most potent of the rocket planes, made its first flights.

Although never accorded the publicity associated with the Mercury or Gemini programmes, the X–15 programme, which continued through to 1968, established a great deal of important data on high-speed flight in the atmosphere vital to the development of re-entering winged vehicles of the Shuttle type. The 23-tonne aircraft, 15 metres long and with a wing-span of 6.8 metres, set some impressive records, including a speed record of 7274 km/hour (nearly seven times the speed of sound), set on 3 October, 1967, and the world altitude record for an aircraft, set on 22 August, 1963, of 108 kilometres; this flight was virtually into space, for all practical purposes.

NASA collaborated with the U.S. Air Force between 1958 and 1963 in the U.S.A.F.'s design study of a small, hypersonic delta-winged vehicle, known as 'Dyna-Soar', to be launched into space by means of the Titan III vehicle and designed to land on a runway. The project was cancelled in 1963. After this NASA joined the U.S.A.F. in examining the possibility of building an 'aerospaceplane' which would land on *and take off from* a runway. Since the early nineteen-fifties tests have been carried out with various designs of 'lifting bodies', wingless craft whose basic shape provides the lift necessary for flight, which have investigated all kinds of questions relating to the aerodynamics of re-entering craft. All of these played a role in specifying the final design of the re-usable Shuttle.

Official backing for the development of a re-usable space transportation system gathered momentum when, in 1967, the President's Science Advisory Committee recommended studies of more economical 'space ferry systems' offering the possibility of recovering all of the vehicle and opening up rescue provision in space. During the period 1969–71 NASA put out study contracts to industry to define possible systems, and the initial proposal which emerged was for *two* piloted vehicles – a booster and an orbiter, both winged vehicles which would glide back through the atmosphere to land on a runway, the orbiter being launched 'piggy-back' on the booster.

Financial constraints forced the abandonment of this rather attractive scheme, and the final design adopted was for a winged orbiter to be launched with the aid of unmanned, but recoverable, boosters. On 5 January, 1972, President Richard M. Nixon gave the Space Shuttle programme his official sanction, remarking that this system would 'transform the space frontier of the 1970s into familiar territory, easily accessible for human endeavour in the 1980s and 1990s'.

The space transportation system, or Shuttle, consists of the orbiter which is fired vertically from a launchpad mounted 'piggy-back' on a giant external fuel tank to which are attached two powerful 'strap-on' solid-propellant boosters. The Orbiter looks like a delta-winged aircraft, having an overall length of 37.25 metres, a wingspan of 23.80 metres, and a maximum height to the top of the tailfin of 17.25 metres; its empty weight is about 75 tonnes. The bulk of the Orbiter is taken up by a cargo bay 18 metres long and 4.5 metres wide, which can accommodate a payload of up to 29.5 tonnes. When in orbit the doors of the cargo bay are opened and satellites can be lifted out of, or into, the bay with the aid of a 15-metre long manipulator arm controlled by the crew. The Orbiter can be used to carry large single payloads, or several smaller payloads simultaneously into orbit; broken-down satellites can be rescued from orbit and brought back to Earth or, if the problem is not too serious, the repair and servicing can be carried out in orbit. Maximum space endurance planned for the orbiter is about 30 days, but missions of a week's duration are more likely to be the norm.

The external tank, which is 46.88 metres long by 8.46 metres wide, contains some 600 tonnes of liquid oxygen and about 100 tonnes of liquid hydrogen to supply the Orbiter's three main engines with propellant during the flight to orbit; each of the Orbiter's main engines achieves a peak thrust of 168 tonnes. The two solid-propellant boosters each provide a peak thrust of over 1300 tonnes, and burn for just over two minutes, dropping off and splashing down with the aid of parachutes when they are expended; they are then towed back to the Kennedy Space Center for refurbishment.

Lift-off of the 56-metre high configuration is achieved with the aid of the two boosters and the three Space Shuttle Main Engines (SSME). Since the launch weight is about 2000 tonnes, but the combined thrust is about 3000 tonnes, the Shuttle accelerates away from the launchpad much more rapidly than, say, the Saturn V. After just over 2 minutes it reaches a height of about 50 kilometres and a speed of 4,600 km/hour, and the boosters drop away. The main engines continue to fire until the Shuttle has reached an altitude of about 120 kilometres and a speed of about 27,000 km/hour some 8.6

minutes from lift-off. They are then shut down, and the fuel tank, now empty, drops away to burn up in the atmosphere (the tank is the only major component which is not re-usable). Two smaller orbital manœuvring engines (OME) are used to place the Orbiter in the required orbit.

To return to Earth, the Orbiter is first turned round so that it is travelling rear-first; the orbital manœuvring engines are fired to slow down the spacecraft and begin its descent to the atmosphere. The Orbiter is then turned again so as to come down forwards, but with its nose tilted up at an angle of about 40° so that the greatest heating occurs on the nose, the bottom, and on the leading edges of the wings. To protect it from the heat of re-entry, the Orbiter is covered by more than 31,000 heat-resistant tiles; as it plunges into the atmosphere at about 28,000 km/hour the external temperature builds up to about 1500°C. After gliding through the atmosphere for a distance of about 7000 kilometres, the Orbiter makes a steep approach to the runway and touches down on its undercarriage at a speed of some 370 km/hour. The original intention was that the Shuttle would be refurbished and ready to go again two weeks after touch-down, but in practice it seems most unlikely that so rapid a turn round will ever be achieved.

When the Shuttle project was approved in 1972, it was envisaged that the first manned orbital flight would take place in the spring of 1978, but in practice the road was by no means a smooth one. As a result of persistent technical problems, in particular, problems with the Shuttle main engines, and with the protective tiles which showed a marked reluctance to stay in place, the launch date slipped again and again, to the extent that, in 1979 and 1980 there was a feeling that the date would always remain about nine months away!

Progress on constructing the Orbiter went well with the first prototype, named 'Enterprise'. It was rolled out of the NASA/Rockwell International Facility on 17 September, 1976, and transferred to NASA's Dryden Flight Research Center at Edwards Air Force Base for approach and landing flight tests. For these tests, carried out in 1977, 'Enterprise' was carried on top of a modified Boeing 747 'Jumbo', purchased second-hand by NASA in 1974 and modified for its new role. Three 'captive' test flights were flown, with the Orbiter attached to the 747 throughout, then five 'free' flights where the Orbiter was released to glide back to base. Test flight crews for these missions were, Fred W. Haise, Jr. with Charles G. Fullerton, and Joseph H. Engle with Richard H. Truly.

Work on the propulsion system proceeded under various pressures. If a Shuttle launch were achieved in 1979 then it might be possible to take a rocket motor in the cargo bay to boost the ailing Skylab into a higher orbit and prevent its premature demise. As 1978 progressed, the target date for the first launching slipped from March 1979, to June then to September. On 27 December, disaster struck – the Shuttle main engine caught fire while on test. In January 1979, it was announced that the first launch had been rescheduled for November. All hope of saving Skylab was abandoned, and it made its spectacular re-entry in July, amid a blaze of adverse publicity. Further engine problems ensued as the year went by; the propulsion system seemed to be jinxed. The recalcitrant tiles continued to give problems as more than half of them had to be rebonded to ensure they did not shake loose in flight. Target dates came and went.

Finally, on 24 November, 1980, the first Orbiter scheduled to go into space – 'Columbia' – was moved from the Orbiter Processing Facility to the giant Vehicle Assembly Building to be mated with the external tank and the solid-propellant boosters. Launch date was set for March 1981.

Minor setbacks pushed the launch date to 10 April, but this time all looked to be going well. The countdown proceeded to within nine minutes of lift-off, then stopped when a malfunction in the onboard back-up computer refused to resolve itself. There was no option but to delay for two days. The huge propellant tank had then to be drained and refilled since the highly volatile propellants could not be stored there for any length of time. Frustration mounted.

On 12 April, 1981 – exactly twenty years after Gagarin's first spaceflight – the day dawned fair for the first flight of the Space Shuttle. Crew for the mission comprised Mission Commander John Young, veteran of two Gemini and two Apollo missions who was returning to space for the first time since 1972, and Pilot Robert Crippen, for whom this was to be his first spaceflight. At T–7 seconds the main engines roared into life. When these had built up to 90 per cent of full power, the solid boosters were ignited, the hold-down clamps released, and at 07.00 Eastern Standard Time, STS–1 – the first Space Shuttle mission was on its way to insertion in a near-circular orbit at an altitude of just over 270 kilometres.

The two days in orbit were spent testing the spacecraft systems, including the operation of the cargo-bay doors. Television cameras revealed some minor damage to the protective tiles; about 16 had come off, but not in a vital area. Nevertheless, there was natural anxiety about this – especially after the difficulties which the tiles had caused earlier in the programme. The historic flight of Columbia had re-awakened the world's dormant interest in manned spaceflight, and excitement mounted as the time for the return to Earth approached.

The payload bay doors were closed and, 53 hours 5 minutes into the mission the retro burn was made to start the descent. For the first time ever a manned, winged vehicle would re-enter from orbit, and at a speed of twenty-five times the speed of sound. The spacecraft entered the inevitable period of radio blackout, as the outside temperature reached 1500°C and the build-up of hot ionized gases rendered radio communication impossible. Seventeen agonizing minutes later, communication was re-established and at 13.21 EST on 14 April, Columbia touched down right on the button on the runaway at Edwards Air Force Base. The cheers of the quarter-million strong crowd symbolized the relief and exultation felt at the successful conclusion of a brave new venture – the first flight by a re-usable vehicle, the first winged vehicle to re-enter from orbit, the first to land on a conventional runway, the first American spacecraft to touch down on land, and the first ever directly witnessed to do so by the general public. After the trials, tribulations, and disappointments, after the delays and setbacks, the first flight of the Shuttle had achieved all that was asked of it.

The return to the Kennedy Space Center was made in rather more sedate fashion on top of the 747, and work proceeded to get Columbia ready for the next big test – the second flight into orbit. Until that had been achieved, the concept of re-usability could not be regarded as proven.

Plate 18. The Space Shuttle 'Columbia' seen on the runway at Edwards Air Force Base, California after landing at the completion of its first orbital flight on 14 April, 1981. (NASA)

Seven months after the first mission, Columbia blasted off for a second time, commanded by Joe H. Engel – who lost the chance of becoming the twelfth man to walk on the Moon when he was pulled out of the Apollo 17 team to allow geologist Harrison Schmitt to go instead – and piloted by Richard H. Truly. For both men, this was to be their first spaceflight. For the first time, Columbia carried a payload, five groups of experiments mounted on a 'pallet' built by British Aerospace as part of the ESA 'Spacelab' project, and on this mission the Canadian-built manipulator arm was tested and found to work well. The Mission was cut short because of a faulty fuel cell, but was nevertheless deemed a success with 90 per cent of the mission objectives attained. Touchdown took place on 14 November, 1981. For the first time in history a spacecraft had flown twice and landed twice. After decades of dreaming and speculation, a re-usable space transportation system was a proven reality.

The third mission – STS–3 – was flown in March 1982 by Mission Commander Jack R. Lousma (veteran of the 59-day second mission to Skylab in 1973) and C. Gordon Fullerton, who had not previously flown in space but who had taken part in three of the 'Enterprise' approach and landing glide flights in 1977. During the scheduled seven-day mission Columbia was put through a wide variety of severe tests, particularly concerned with the effects of prolonged heating and cooling at different attitudes relative to the Sun. With the Edwards landing site waterlogged and out of action, the crew had to remain for an extra day in orbit because of a sandstorm at the alternative White Sands landing site. When that had subsided, Columbia came safely to touchdown there on 30 March.

After one more proving flight, Columbia is expected to begin operational service in the autumn of 1982.

The Space Shuttle cost 8.8 billion dollars to develop, a figure which escalated well over target as delays and complications mounted. One of the casualties on the way was the planned 'Space Tug', a re-usable vehicle to manœuvre payloads from the Orbiter to higher orbits, if required. Each Orbiter costs about 300 million dollars to build but, because it is hoped to use each one at least one hundred times, the cost per payload into space should be significantly less than the cost involved using ordinary launch vehicles. However, it must be admitted that the actual saving achieved will not be as much as had been imagined when the Shuttle was first conceived.

Each Shuttle flight costs 30-35 million dollars (1981 prices), but small packages carried in empty corners of the cargo bay can be taken up into space cheaply. These 'Getaway Specials' can be launched for only about 100 dollars per kilogram provided they fit in a standard container. When the first American satellite, Explorer 1, was placed in orbit in 1958, the cost of getting it up there worked out at around 200,000 dollars per kilogram. Looked at in these terms a Getaway Special represents very good value indeed!

At present NASA has payloads lined up for the first fifty operational flights, and among these are Space Telescope (of which more will be said in Chapter 12), and Spacelab which, as its name suggests, is a manned laboratory to be taken into space in the cargo bay, and returned to Earth for refurbishing and equipping with new experiments. Built by ESA, Spacelab consists of two basic elements: a pressurized laboratory in which up to four scientists can work in a 'shirt-sleeve' environment, and an external 'pallet' – a platform which carries instruments which require direct exposure to space. Other possible permutations include a full-sized laboratory (with no pallet) and a full-sized pallet operated from within the Orbiter's crew compartment. The first Spacelab was delivered to NASA in December 1981 and is scheduled to make its first orbital mission in September 1983.

A major disadvantage of the Shuttle is that it cannot reach altitudes above about 800 kilometres and so cannot, for example, reach the vitally important 35,800-kilometre altitude geosynchronous orbit – the prime orbit for communications satellites. To place satellites in higher orbits, or to launch probes into interplanetary space, an upper stage launcher must be attached to these payloads. A number of proven vehicles can fit into the cargo bay – the Centaur, for example – and NASA envisaged using such vehicles as an 'interim' upper stage while development work proceeded on a three-stage 'inertial upper stage' launcher. These plans have been hit by financial cut-backs, and the two-stage version of the 'inertial upper stage' is all that is likely to be available in the near future. The fact remains that to make full use of the flexibility offered by the Shuttle, the means must be available to launch interplanetary probes and to place satellites readily in the required orbits. The development of a genuine 'Space Tug' must come before much longer if NASA is to make best use of its Space Transportation System.

The Shuttle will certainly not have a monopoly so far as launching commercial payloads is concerned. Other nations pressing ahead with the development of independent launchers include India, Japan, the People's Republic of China, and, of course, the European Space Agency. ESA's Ariane offers serious competition to the Shuttle for one-off launches, especially for placing communications satellites in the all-important geostationary orbit. In its present form it can place a 1700-kilogram payload into a transfer orbit with apogee at around 36,000 kilometres; a motor attached to the payload then fires to kick the satellite into a circular orbit, and nudges it along to the required position.

Uprated versions of Ariane currently under development should increase this capability considerably during the nineteen-eighties: Ariane 3 should cope with placing the 2.4-tonne direct communications satellite, L–Sat, in orbit in 1986, and possible uprated versions 4 and 5 could increase the payload placed in transfer orbits to 36,000-kilometre altitude (geostationary transfer orbit – 'GTO') to 3.5 and 5.5 tonnes respectively. If given the go-ahead, Ariane 5 might be able to carry a small European Shuttle-type craft in the nineteen-nineties.

The Ariane 1 vehicle, 47 metres long, and with an all-up weight of 208 tonnes, is a three-stage vehicle generating a first-stage thrust of 245 tonnes. It made its first flight from Kourou, French Guiana, on 24 December, 1979, and on that occasion successfully placed a test satellite in orbit. The second launch was less auspicious, a failure in the second stage leading to the break-up of the launcher, which fell into the Atlantic Ocean together with its payload of two West German satellites. After lengthy investigations, the problem was solved and the third and fourth launches have been completely successful. The third launch carried three satellites, two of which – as intended – were placed in geostationary orbit; these were Metosat 2, a European weather satellite, and 'Apple', an experimental Indian communications satellite. The fourth and final test flight took place on 20 December, 1981 and placed the 'Marecs' maritime communications satellite in geostationary orbit.

With its test programme behind it, Ariane already has some 30 satellites tentatively booked for launching up to mid-1985. Space clearly is becoming a competitive commercial field where rival launching agencies will have to offer the best possible service if they are to increase, or even maintain, their share of the lucrative satellite market. The commercial potential is highlighted by the announcement early in 1982 that a 'Space Transportation Corporation' has been formed which is confident of raising a billion dollars to make a private purchase of a fifth Shuttle Orbiter to complement the activities of the four which NASA itself currently plans to bring into operation.

But what of Soviet intentions in the field of re-usable spacecraft? The Soviet space programme remains shrouded in a web of secrecy, but there is no doubt that Russia is working to develop a Shuttle-type vehicle, although the nature and time-scale of this venture are uncertain. Reports indicate that work is progressing on a vehicle considerably smaller than the American Shuttle, a delta-wing vehicle similar to the one-man 'Dyna-Soar' which was cancelled in 1963. It has allegedly been seen undergoing drop tests from a Tu–95 'Bear' bomber, and indications of the construction of a new runway at Tyuratam have been taken to imply that it is planned to land there.

On the face of the available evidence it seems likely that the first Soviet shuttle will be a small vehicle, essentially a personnel carrier to transfer crews to and from space stations. When it will fly is anybody's guess, but it may be within the next year or two. There are rumours, too, of a much more massive heavy-lift re-usable vehicle, but tangible evidence is lacking. All will eventually be made clear by the course of events.

What is certain is that the Shuttle has ushered in a new era of space *exploitation* – an era of growing activity in near-Earth space when launch and recovery operations will be a matter of everyday routine, and commercial involvement will be high. It has also opened the doors to some longer-range prospects, too, but these are the subject of the next chapter.

12

The Way Ahead

How will our activity in space develop over the next twenty-five years? The probable trends in the next ten years are relatively easy to foresee, but beyond that reasoned prediction rapidly gives way to speculation and, mindful of some of the wilder predictions which have been made during the past few decades, we should be wary of being too dogmatic about the future.

In the satellite field, the way ahead, at least in the 'eighties, seems quite clear-cut. Applications satellites will increase in numbers, sophistication, and range of activities. The biggest step forward in the communications satellite field will come with direct broadcasting to individual households from high-powered geostationary satellites. Operational satellites of this kind will become available in the latter part of this decade, and as we move into the nineteen-nineties, the range of programmes, communications, and information services which will come directly into every home will expand to a bewildering extent.

Already we are beginning to experience the effects of an unprecedented communications revolution with video recorders, video games, home computers, and information services available on domestic television sets at the touch of a button. Already we have become used to the fact that we can dial direct to almost any part of the globe and hold a telephone conversation with antipodean friends with the same degree of clarity (or the lack of it!) as we experience when making a local call. Communications satellites will expand and enhance all these facilities, making true global communication – in its widest sense – a reality. The day is not far removed when the global personal communicator will become available, an instrument no larger than a pocket calculator will provide the means to communicate, via direct satellite links, to anyone, anywhere, on the face of this planet, and to 'plug in' to any desired information source. Conversely, the instrument could be used to locate the position of any individual on the Earth's surface – not altogether an appealing prospect – but, within the 25-year timescale, an almost inevitable consequence of the communications explosion.

Earth surveillance and monitoring satellites of all kinds (meteorological, oceanographic, Earth resources) will play an increasing role in the economic life of our planet, and commercial organizations will play an increasing role in the construction, launching, operation, and maintenance of these facilities. More and more nations will become involved in space activity, and band together to form viable groupings for this purpose. Just as the individual European nations have pooled their resources to form the European Space Agency so it seems, that – as India has advocated – the less-developed nations may come together to form a 'Third World' Space Agency. At all events, satellites will play an increasingly significant role for *everyone* on the planet Earth.

The immediate outlook for the exploration of the Solar System is mixed.

So far as NASA is concerned the escalating costs of developing the Space Shuttle, combined with the heavy pruning of its budget by successive American administrations, has all but annihilated its planetary exploration programme. Apart from the ongoing Voyager mission (which will take Voyager 2 to Uranus in 1986 and, if it is still in commission by then, to a fly-by of Neptune in 1989) the only survivor from the exciting range of projects which NASA had in the planning stage is the 'Galileo' mission – a plan to send an atmospheric probe and an orbiter to the giant planet Jupiter.

Galileo is an exciting mission, of that there is no doubt, but NASA had to fight a desperate rearguard action to preserve it. Originally the intention had been to use a three-stage 'inertial upper stage' (IUS) vehicle taken into orbit by the Shuttle, to launch the probes towards Jupiter. With delays in the Shuttle programme, the planned launch date slipped from 1982 to 1984, and then to 1985 using a Centaur upper stage instead of the IUS. The whole mission was under threat of cancellation late in 1981, but has survived, in somewhat emaciated form, and now seems certain to go ahead, but with a two-stage IUS instead of the more powerful Centaur. This latest cut in the capability of the launch vehicle implies that the spacecraft cannot go direct to Jupiter (the two-stage IUS is not sufficiently powerful) and must instead follow a tortuous trajectory, making use of gravitational 'slingshots' from both Earth and Mars to reach its target, two years later than planned.

In order to preserve Galileo in the face of drastic budgetary cuts (nearly 500 million dollars in 1982 alone), NASA has had to pull out of two major Solar System missions, Venus Orbiting Imaging Radar (VOIR) and the International Solar Polar Mission (ISPM). VOIR, as its name suggests, was intended as a Venus orbiting mission using high-resolution imaging radar to improve on the Venus mapping carried out by Pioneer Venus after its arrival in orbit round the planet in 1978. The ISPM mission was to have been a joint venture with ESA, each space agency supplying one spacecraft, for a mission of the utmost scientific interest, to investigate the Sun and interplanetary space. Launched via the Shuttle, the two spacecraft, which were to have had different but complementary instrumentation, would have travelled out to Jupiter and used Jupiter's gravitational field to hurl them out of the ecliptic (the plane of the planetary orbits) into trajectories which would take them directly over the Sun's pole allowing them to study an important region of the Sun which cannot readily be studied from Earth. While in flight, they would have investigated the nature of the interplanetary medium – particles and magnetic fields – far above the region of space so far explored by spacecraft which, up to now, have confined themselves mainly to the orbital planes of the planets.

Almost certainly, ESA will go-it-alone with one spacecraft using a Shuttle launch in 1986, but the value of the mission will be seriously reduced because the comparison of data between the two spacecraft will no longer be possible. ESA is most annoyed, but if NASA were to preserve Galileo, it had no option but to pull out of ISMP.

For NASA, the 1970s was the decade of the planets, a period in which they dominated the exploration of the Solar System, and brought us a wealth of new knowledge and an abundance of stunning photographs of Mercury, Venus, Mars, Jupiter, and Saturn. The 1980s will be a decade of retrenchment, 'wilderness years' by comparison with the heady days of the 'seventies. Even so, if all goes well, there will be exciting results to come from Jupiter, Uranus, and Neptune, and in the longer term, it is likely that NASA's fortunes will revive, and the Agency's involvement in planetary exploration will once again become very substantial.

There can be little doubt that the Soviet Union will carry out further planetary exploration during the remainder of the 'eighties, but the only missions for which prior information is available are Vega 1 and Vega 2, scheduled for launching in 1984. Each spacecraft will deliver a French-built instrumented balloon intended to float in the Venusian atmosphere to relay information about atmospheric conditions and wind speeds. Both vehicles will then continue past the planet to make a close encounter with Halley's comet in 1986.

ESA has a number of interesting projects in mind, apart from the ISPM. The 'Giotto' spacecraft, currently under construction, and which is scheduled to be launched by Ariane in July 1985, is intended to fly through the tail of Halley's comet in March 1986, shortly after the comet's closest approach to the Sun. Heavily shielded to protect it from impacts by the dusty particles of the coma and tail, the spacecraft hopefully will pass by the nucleus at a range of about 1000 kilometres, sending back photographs and analyses.

Other missions which seem likely to be undertaken within the next decade or so include the sending of remote-controlled roving vehicles to Mars and, perhaps, Venus, these being followed by a remote sample return mission to Mars. An asteroid fly-by mission also seems likely, and by judicious selection of a trajectory, several different asteroids could be inspected from close range by just one spacecraft. Both NASA and ESA have produced design studies for this sort of mission.

Looking a little further ahead, there is a multitude of possible missions, many of which must come to fruition well within the next twenty-five years – orbiters and atmospheric probes to all of the giant planets, a spacecraft to investigate remote Pluto and its strange satellite, orbiters and landers to some of the more interesting planetary satellites and, of course, an atmospheric probe to Saturn's major satellite, Titan. Comets will be further explored by missions which match orbits and achieve rendezvous, allowing prolonged investigation, rather than the all too brief fly-through which Giotto is to attempt. Following the suggestion by Professors Hoyle and Wickramasinghe that micro-organisms may exist in comets, the results of such missions should be particularly fascinating. Asteroid rendezvous missions (it is difficult to talk about 'landing' on the smaller asteroids which have practically negligible gravitational fields) and surface samplers will give us an indication of the potential mineral wealth available to future 'asteroid miners'. The 'solar plunger' mission may be undertaken, the idea being to use Jupiter's gravity to swing a spacecraft on to a path which will drop it straight into the Sun; ESA studies suggest that the probe could continue to relay information until it reached a distance of about two million kilometres above the incandescent surface of our parent star. By the first decade of the twenty-first century, we can be confident that every major body in the Solar System, and a good selection of the minor ones, will have been visited by at least one unmanned spacecraft.

Our exploration of the remote recesses of the universe will continue with the aid of orbiting instruments in satellites and space stations. Most exciting of the currently planned missions is the Space Telescope, to be placed in a 600 kilometre-high orbit by the Shuttle, hopefully in 1986. With a 2.4-metre aperture mirror – which was completed in 1981 – it will be by far the largest telescope to be placed in space and, with its sophisticated electronics, will be able to detect stars and galaxies about fifty times fainter than the faintest attainable by any existing ground-based telescope. As a result it will be able to probe to much greater distances and may provide us with crucial information on the origin and future evolution of the universe as well as on other vexed questions, such as the formation of galaxies.

Other instruments and satellites will continue the exploration of the universe at gamma-, x-ray, ultraviolet, and infrared wavelengths. Large space-based radio telescopes are also a possibility, Soviet astronomers having proposed the construction of instruments of this kind with apertures of between 1 and 10 kilometres. Less spectacular, perhaps, but fundamental to constructing an accurate picture of the universe, is astrometry – the science of measuring the positions, motions, and distances of stars; ESA has prepared plans for an astrometric satellite, named 'Hipparchos' (after the great Greek astronomer who, in the second century B.C., made fundamental observations of the positions and brightnesses of stars), which should extend by a factor of between 10 and 100 both the accuracy of these types of observation and the number of stars investigated.

Manned spaceflight in the nineteen-eighties is likely to be confined to Earth-orbital missions – although it is always possible that the Soviet Union will suddenly announce a dramatic long-range mission. The exploitation of space will be the primary aim in the immediate future. Just as the pioneering sea voyages of Columbus, Vasco da Gama, Cook, and others, were followed by periods of colonization, settlement, and trade, so we are moving into an era when mankind – having invested the resources in a period of exploration – will expect to receive a tangible return from its investment. A major harvest is already being reaped, of course, with communications and earth-monitoring satellites, but the next major step, involving manned participation, will be space industrialization, and this will require the construction of substantial, long-endurance space stations.

Drawing on experience already gained with Skylab, and experience which will be gained with Spacelab, it seems likely that NASA and ESA will move towards the construction of a space station capable of housing a crew of 6 to 12 by the late nineteen-eighties. One scheme which has been proposed is to use the Shuttle external propellant tank as the basis of a large, single unit, station to be fitted out in orbit – an extension of the earlier 'wet' laboratory concept which pre-dated Skylab. An alternative, and more probable, approach is to build up a space station in orbit from a number of standard-sized modules brought up in the cargo bay of the Shuttle. Bearing in mind the size of the cargo bay, the maximum size for each module would be about 15 metres by 4 metres. Quite probably, the space station would be built from a number of Spacelab units.

The Soviet Union, already having a decade of experience with the Salyut programme, seems likely to follow a similar approach, assembling a modular space station out of a number of Salyut-sized units. This development may take place very soon.

Among the beneficial activities which will be pursued are the construction of space 'factories' to undertake large-scale materials processing. Skylab and Salyut experiments have already demonstrated that certain materials can be made better and purer in space, and others can be made which cannot be produced at all on Earth. Space materials processing is likely to become a significant factor in the world's economy by the turn of the century. Another of the first major developments is likely to be the construction of satellite solar-power stations (SPS) which would collect solar energy and beam it down to the Earth in the form of microwaves (which would pass through cloud) to receiving stations which would turn the microwave radiation back into electrical power.

This is a serious proposition which is being actively pursued in feasibility studies in the United States at the present time. As presently envisaged, a typical operational SPS would measure about 16 kilometres by 4 and would transmit from geostationary orbit something like 5000 megawatts of power – enough to supply about a million homes. Solar power delivered in this way would be largely pollution-free (although the effects of the microwave beams on the atmosphere and ionosphere would have to be studied) and, of course, the Sun should remain a constant source for at least the next five billion years. A smaller scale test SPS could be assembled and operating by the mid 1990s, and the first full-scale system could be supplying its contribution to the United States national grid by the year 2000. With dwindling supplies of non-renewable resources such as oil, SPS systems could provide a significant contribution, at an economic cost, to global energy requirements during the twenty-first century and, at the very least, could help buy the time necessary to develop operational fusion-power stations (producing energy by essentially the same process that goes on in the solar interior) here on Earth.

Other, more modest, projects, include designing solar reflectors to light cities by night.

Projects of this kind will require the transport of large numbers of men (engineers and construction workers) and masses of materials into orbit and this will demand a space transportation system with a greater payload capacity than the present Shuttle. One possibility for a heavy-lift launch vehicle (HLLV) for materials alone, is to use Shuttle-type boosters without the manned orbiter, another is to develop an unmanned re-usable vehicle. Alternatively we may move towards the piloted booster/piloted orbiter concept which was originally proposed prior to the present Shuttle, but uprated to take payloads of over 100 tonnes into low Earth orbit. Yet another concept is the aerospace plane, a winged vehicle which would take off from a runway using jet engines, then accelerate to several times the speed of sound before cutting in its rocket motors to blast into orbit. After re-entry it would be able to fly, manœuvre, and land like a conventional aircraft. A vehicle of this kind is unlikely to make its appearance before the turn of the century, but it would be very unwise to be dogmatic.

A substantial orbital transfer system (OTS) capable of taking these large payloads to the SPS assembly site in geosynchronous orbit would also be essential, and we are almost certain to see developments of this kind before the 1990s are far advanced.

An even more ambitious scheme is the building of large, permanently inhabited space colonies capable of housing thousands, or even millions of people. Serious consideration of this possibility began with feasibility studies carried out by Professor Gerard K. O'Neill of Princeton University in the early 1970s, although the eventual construction of such colonies had been envisaged by Tsiolkovsky and others many decades previously. These structures might take the form of spheres or cylinders (possibly several kilometres long) and would contain air, soil, lakes, and houses, as well as associated industrial and agricultural units. All the power requirements would be provided by the Sun, and artificial gravity would be produced by spinning the 'colony' on its axis.

How and where would such structures be built? O'Neill suggested that the best positions would be in the orbit of the Moon but at positions 60° ahead and 60° behind the Moon itself. Placed at these points, the colonies would be in stable orbits and would remain in these same positions relative to Earth and Moon as they travelled round in their orbits (these points are known as 'Lagrangian points' and are designated 'L–4' and L–5'; flourishing L–5 societies now exist to promote this aim). The materials to build these giant structures would be mined from the Moon as it requires far less energy to take material from the lunar surface into orbit round the Earth than it does to fetch it up from the Earth's surface.

Whether or not such large-scale structures are built in the foreseeable future, it is entirely reasonable to suppose that lunar minerals will be used, within a few decades, to build structures in Earth orbit for the very good reasons that it would be cheaper, would not deplete the resources of the Earth, and would minimize terrestrial pollution.

Manned missions to the Moon are likely to be resumed in the 1990s, with the intention of establishing a base for scientific observations and as a precursor to future mining and constructional activities. An observatory for radio astronomy – shielded from terrestrial radio interference – may well be constructed within 25 years. It is very likely that the next men to set foot on the Moon will be Russians, who may well undertake a manned lunar mission before 1990.

Manned missions to Mars by about the year 2000 are a distinct possibility. A NASA design study was produced in the 1960s for a twelve-man mission to Mars using two vehicles with nuclear-powered stages to accomplish the trip. Again, it may well be that the Soviet Union will be the first nation to accomplish this mission. It would be nice to think, however, that when this next great leap in manned space exploration occurs, it will be undertaken as an international mission.

New forms of propulsion will be required if we are to cut down on flight times for interplanetary missions – manned or unmanned. Two possibilities which have already seen some development work are the nuclear rocket and the ion rocket. Studies of the nuclear rocket began in the United States in 1958 and led to the development of NERVA (Nuclear Engine for Rocket Vehicle Application) in which a nuclear-power plant was used to heat liquid hydrogen fuel to exhaust velocities several times higher than the best attainable by conventional rockets; a thrust of about 24 tonnes was developed. Ion rockets, utilizing electric fields to accelerate charged particles and so produce thrust have already been used as experimental onboard thrusters.

The Solar Electrical Propulsion System (SEPS) has been examined in detail by NASA. Using large arrays of solar panels to provide the required electrical power, spacecraft propelled by this means could accelerate gently but steadily to speeds several times higher than the best that has yet been attained by existing rockets. The potential of such systems for interplanetary missions is very great – as would be their physical size! One system which was being actively considered by NASA (prior to the budget cuts) for missions in the 1980s would have required 'wings' of solar panels about 150 metres long.

For missions into the outer reaches of the solar system, where solar radiation is weak, nuclear-powered ion rockets may be used.

If all goes well, and if we survive the next few crucial decades, mankind may well have travelled the length and breadth of the Solar System by the end of the twenty-first century, and – in time – more human activity may be taking place in space than on the surface of our home planet. Before too long, Man may turn his attention to the next big challenge– interstellar flight. To reach even the nearest star within an acceptable period of time (a few decades) a spacecraft would need to travel at more than 10 per cent of the speed of light. Since present-day launchers can barely attain 0.005 per cent of light velocity, to attain the velocity necessary to send a probe to a nearby star within the span of a human lifetime would seem to be out of the question in the foreseeable future. Yet, this may not be so. Fusion rockets should be capable of attaining exhaust velocities of thousands of times greater than those attainable by present-day liquid-fuelled vehicles – and fusion technology should be with us early next century. The British Interplanetary Society, in its far-sighted *Project Daedalus* report, suggested that the key to interstellar flight would be the nuclear-pulse rocket powered by a series of tiny thermonuclear explosions produced by imploding tiny fuel pellets; about 250 detonations per minute were envisaged for this ambitious vehicle intended to send a 500-tonne payload on a 50-year flight towards Barnard's star, some six light years away. Such a mode of propulsion relies only on an extension of existing technology; the implosion of fuel pellets as a means of producing fusion reactions is already being studied at a number of laboratories throughout the world.

Such exciting prospects, however, lie well beyond the next 25 years. Between now and the fiftieth anniversary of Sputnik 1 we shall see a massive increase in space activity in Earth orbit, both in the form of unmanned satellites and in the form of manned space stations, manufacturing facilities and, perhaps Solar Power Stations. The era of space exploitation will be with us. Unmanned probes will explore the entire Solar System and Man will resume his personal explorations with missions to the Moon and possibly to Mars.

The prospects are limitless, provided we can grasp the opportunities. The dark prospect which lurks before us is that space becomes the new arena for armed conflict. If anti-satellite systems are further developed, then an arms race in space may ensue. The Shuttle itself has definite military capabilities, for direct reconnaissance, for the deployment and retrieval of military satellites, and for monitoring, disabling, or even 'capturing' opponent's satellites. High-powered laser weapons based on space platforms have been mooted as a means of destroying ICBMs in flight, shortly after lift-off. Many military possibilities exist, and the prospect of 'war in space' is no longer a fanciful one.

The next 25 years will be crucial, not only for space exploration, but for the whole future of the human race. The prospects for unlimited progress in space are there. The choice is ours.

13

Recent Developments – April 1982 to September 1984

The period from April 1982 to September 1984 was an eventful one, particularly in the field of manned spaceflight. Both America and the Soviet Union had successes and setbacks, but both nations demonstrated a growing ability to carry out repair and servicing operations in space and to rescue space missions from potential failure.

Salyut/Soyuz

The Soviet Union continued its space station programme with Salyut 7, which included a number of improvements aimed at providing a more comfortable working environment. On 14 May, 1982, cosmonauts Anatoli Berezovoy and Valentin Lekedev docked their Soyuz T–5 spacecraft with the space station to commence a record-breaking 211-day stay. During their time aboard, their supplies were replenished by four Progress 'freighters', and they were visited by two short-stay crews.

The first visiting party included Lieutenant Colonel Jean Loup Crétien, the first French astronaut, while the second aroused widespread interest because it included Svetlana Savitskaya, the first female cosmonaut since Valentina Tereshkova and only the second woman ever to fly in space. The prime role of her mission was to determine the physiological effects of zero gravity on a woman, but some Western correspondents felt that the mission was also intended to upstage the forthcoming flight of the first American woman astronaut, Sally K. Ride, on the seventh Shuttle mission. Commenting on the decision to send a woman to Salyut, the Soviet news agency, Tass, remarked, 'The presence of a woman exerts an ennobling influence on the microclimate of the small group'. Svetlana Savitskaya is reported to have countered any suggestions that she might play a domestic role aboard the station by commenting, 'Housekeeping details are the responsibility of the host cosmonauts'.

Berezovoy and Lekedev's mission covered a wide range of activities from materials processing (1.5 kg of ultra-pure semiconductor crystals were returned to Earth) to tests of techniques for overcoming space sickness. During an EVA lasting just over two and a half hours, Lebedev tested techniques for assembling large structures in space. The mission ended with a hazardous night landing during a snowstorm on 10 December, 1982.

On 10 March, 1983, Cosmos 1443, a module comparable in weight with the space station, docked with Salyut 7, and on 20 April three cosmonauts were dispatched towards the station in Soyuz T–8. They were unable to dock and returned to Earth two days later. In view of the prior publicity accorded to the mission, the outcome was a disappointing setback to the programme. However, on 27 June, Vladimir Lyakhov and Alexander Alexandrov successfully docked and entered the space station. They unloaded material from Cosmos 1443, which later undocked and returned to Earth a separate capsule containing about 500 kg of cargo from Salyut 7.

A serious crisis arose on 9 September when a major leak developed in the propellant system and the contents of two out of three oxidant tanks were lost into space. This severely restricted the manoeuvrability of the station. Furthermore, the electrical supply from its solar panels had degraded significantly. Salyut's problems were compounded by a dramatic launch pad accident in which cosmonauts Titov and Strekalov were fired clear by escape rockets seconds before the booster that was to have taken them to Salyut – presumably to assist with repairs – exploded.

Despite this accident, some progress was made. In October the usable tank was refuelled by Progress 18, and in early November the long-stay crew made two EVAs to install additional solar panels, using materials sent up in the Progress freighter. Having taken the first steps in rehabilitating the station, Alexandrov and Lyakhov returned to Earth on 28 November.

The next mission to Salyut 7 was launched on 8 February, 1984, and on the following day Vladimir Solovyov, Leonid Kizim and Oleg Atkov docked their Soyuz T–10 craft and entered the space station to begin a mission that was to set yet another new space endurance record. Following a refuelling operation from a Progress freighter, Salyut was boosted to a higher orbit, and early in April the long-stay crew was joined for a week by the three-man crew of Soyuz T–11, a crew that included Rakesh Sharma, the first Indian astronaut.

During late April and May three Progress craft docked with materials and stores, and Kizim and Solovyov made a total of five EVAs, each averaging over three and one-half hours, to carry out vital repairs to the propulsion system. The long-stay crew were again joined by three visiting cosmonauts in July, including Svetlana Savitskaya making her second spaceflight. Savitskaya and Dzanibekov made an EVA of their own to try out metal-cutting techniques and, before their return to Earth, instructed the long-stay crew in techniques that they later used during a 5-hour EVA on 8 August to bypass the propellant leak and seal off the fractured line. The arduous series of EVAs that restored Salyut 7 to full operational capability accumulated more EVA time than the entire Soviet space programme from 1961 to 1983 inclusive, and clearly demonstrated the Soviets' ability to overcome and repair major hardware problems in space.

On 6 September, 1984, Kizim, Solovyov, and Atkov surpassed the previous space endurance record of 211 days.

Space Shuttles

The United States Space Shuttle continued to dominate the space headlines. STS–4, the fourth and final proving flight before the Shuttle entered operational service, commenced with Columbia's lift-off from Kennedy Space Centre on 27 June 1982. Although the launch went well, the two boosters – which were to have been retrieved and refurbished for use on the STS–10 mission – sank to the bottom of the Atlantic Ocean after a faulty switch released their parachutes too soon. Mission Commander Thomas K. Mattingley and Pilot Henry W. Hartsfield flew the orbiter for prolonged periods in various attitudes to assess the effects of solar heating and cooling, further tested the manipulator arm, and launched a Department of Defense satellite, which was only partially successful as the lid failed to come off its infrared sensing telescope, despite judicious prodding by the manipulator arm. The first 'Getaway Special' flew carrying nine experiments from the University of Utah. On 4 July Columbia landed at Edwards Air Force Base, California, touching down for the first time on a concrete runway rather than the dried lake bed.

The fifth orbital mission was launched on 11 November and notched up several 'firsts'. It was the first to deploy two commercial communications satellites, and the first occasion on which the Payload Assist Module (PAM) – a solid propellant motor—was used to project satellites from the low Shuttle orbit to the high geosynchronous one. It was also the first occasion on which four men were launched in a single spacecraft. The major disappointment of the mission was the cancellation of planned spacewalks by Dr. William B. Lenoir and Dr. Joseph Allen due to separate faults in their new, more flexible, space suits. The oxygen circulating fan in Allen's proved troublesome, and Lenoir's did not attain the required pressure because two tiny locking devices, each about the size of a grain of rice, had been left out during the assembly of the million-dollar suit.

After touchdown on 16 November, Columbia was taken out of service to undergo modifications in preparation for the first Spacelab mission. Columbia's mantle was taken on by 'Challenger', the second – and more sophisticated –

orbiter. Challenger incorporated a number of changes, including the addition of a head-up display on the cockpit windows and the removal of the ejection seats. Improvements were made to the tiles, and about 600 of them were replaced by a blanket-like insulation material. An overall weight saving of more than a tonne was achieved in the orbiter itself, and total weight savings on the orbiter, external tank and boosters amounted to over 9 tonnes. A 4 per cent increase was achieved in the power of the orbiter engines.

After some ten weeks of delay due to a cracked fuel line, engine replacements, and contamination of the payload, Challenger – commanded by Paul J. Weitz – achieved a flawless launch on 4 April, 1983. The main payload was TDRS-A, the first of three Tracking and Data Relay Satellites that are to be established in geostationary orbit to relay information from orbiting satellites and spacecraft to a central ground station at White Sands, New Mexico. The existing Earth-based Space Tracking and Data Network can only maintain contact with a low-level satellite for a small proportion of its total orbital period (15 per cent in the case of the Shuttle), but from the vantage point of geostationary orbit the TDRS system should maintain contact for practically all of the time with up to 26 satellites simultaneously. TDRS-A, together with the two-stage solid propellant Inertial Upper Stage (IUS) weighed 18.5 tonnes and was the heaviest payload to be deployed from an American vehicle since the days of Saturn 5.

A malfunction in the second stage of the IUS resulted in failure to circularise the orbit at the correct altitude, but all was not lost, for as a result of a carefully computerised series of firings of the spacecraft's own thrusters over a period of 59 days, the correct orbit was eventually achieved and the spacecraft was nudged along to the appropriate point over the equator in time to provide vital communications support for Spacelab in November. Despite this eventual success the failure of the IUS second stage was a serious setback, as this propulsion system is intended to provide the means to elevate heavy payloads to higher orbits.

Challenger returned to space on 18 June, 1983. Commanded by Robert L. Crippen, STS–7 was the first occasion on which a single spacecraft had been launched with a crew of five. Included in the crew was Sally K. Ride, the first American woman to fly in space. Two communications satellites were deployed and the Shuttle Pallet Satellite (SPAS) – a reusable platform for carrying experimental packages – was for the first time operated outside the orbiter as a free-flying spacecraft and then retrieved and returned to Earth. Inside the orbiter's mid-deck cabin, experimental plants were operated to produce a pure pharmaceutical product and to manufacture precisely identical small rubber spheres suitable for calibration purposes. Space-processed spheres of this kind are expected to be marketed commercially in 1985.

Challenger's third orbital trip, and the eighth Shuttle mission, was commanded by Richard H. Truly and commenced on 30 August. Originally scheduled to launch the second TDRS, it was unable to take this payload because of the problems that had arisen with the IUS. Instead it delivered India's INSAT 1–B to orbit and carried out assorted tests. The Mission also featured the first night launch and landing.

The refurbished orbiter Columbia returned to space on the ninth Shuttle mission on 28 November, 1983. This mission carried the ESA Spacelab into orbit for the first time. The six-man crew – the largest up to that time – was commanded by John Young, commander of the first Shuttle flight, and included two payload specialists (professional scientists with direct responsibility for the onboard experiments): Dr. Byron Lichtenberger and Dr. Ulf Merbold from West Germany, the first non-American to fly in a U. S. spacecraft. During the ten-day mission a vast range of experimental work was carried out on such diverse subjects as materials processing, life sciences, plasma physics, atmospheric physics, and astronomy. Some progress was made in learning how to overcome space sickness – by taking seasickness-type drugs and by minimizing disorientation – and it was shown that the manual dexterity of the crew was significantly reduced in weightless conditions.

The tenth Shuttle mission – designated 41–B and commanded by Vance Brand – took place between 3 and 11 February, 1984. It was a mission of very mixed fortunes. Two communications satellites – Western Union's Westar 6 and Indonesia's Palapa-B – were rendered useless by being left in low orbits after the PAM motors had failed. Two successive failures of this motor – which previously had a good track record – combined with the IUS failure the previous year, dented the Shuttle's image as a reliable space transportation system, although in none of these cases was the Shuttle itself to blame.

However, dramatic success was achieved with the first free-flying tests of the 'Buck Rogers' Manned Manoeuvring Units (MMU) – personal strap-on rocket propulsion systems. First to try out the unit was Navy Captain Bruce McCandless. After flying within the confines of the cargo bay, he set forth on an untethered flight to a distance of some 45 metres from Challenger, and later went out further, to nearly 100 metres' range. Army Colonel Robert L. Stewart followed with a free flight to a similar distance. Both made practice dockings to try out techniques due to be used in a satellite rescue attempt on the next Shuttle mission. As McCandless undocked his MMU to begin man's first-ever untethered EVA, he is reported to have remarked, 'That may have been one small step for Neil (Armstrong) but it was one heck of a big leap for me'.

Challenger made the first landing at the new 4.6 km runway at Kennedy Space Centre, Florida. By cutting out the delays previously involved in ferrying orbiters back from California, Challenger was readied for space again in record time and set off once more, under the command of Robert Crippen, on 6 April. The aim of the eleventh Shuttle mission was to rescue and repair the defunct Solar Maximum satellite (see page 100), which was spinning out of control.

On 8 April, George Nelson set out with his MMU to attempt the hazardous task of docking with the satellite and using his thrusters to halt its spin. The mechanism failed to dock, and an attempt to stabilise the satellite by grabbing a solar panel was equally unsuccessful. The spin was eventually brought under control by means of an onboard magnetic system, but as a result the satellite's power supplies were dangerously depleted. On 10 April, in a remarkable display of precision manoeuvring, Crippen brought Challenger sufficiently close to hook Solar Max with the manipulator arm and bring it into the cargo bay. The following day, in an EVA lasting 7 hours 18 minutes, Nelson and James Van Hoften replaced the failed attitude control module and carried out major surgery on the electronics unit for the satellite's coronagraph. After a period of testing, the satellite was redeployed in a fully operational condition, and it is confidently expected to function efficiently until about 1990.

The success of this operation suggests that servicing, repair and refueling of satellites in space will become an important and cost-effective aspect of manned spaceflight in the years to come. It has also encouraged NASA to undertake a rescue attempt on the Westar 6 and Palapa-B satellites, which failed to leave low orbit in February 1984. Indeed, NASA came under pressure from the insurance underwriters to make such an attempt.

The first launch of Discovery – the third orbiter – was dogged by a succession of mishaps and delays. In the most dramatic of these, two engines ignited and shut down just four seconds before lift-off when the main fuel valve in the third engine failed to open. However, the twelfth Shuttle mission, commanded by Henry W. Hartsfield, eventually got away to a clean launch on 30 August 1984. The crew of six included the second American woman astronaut, Judith D. Resnick. The lift-off weight of 2050 tonnes was the heaviest since the days of Saturn 5, and the 21.6 tonnes of cargo set a new Shuttle record.

During the mission three communications satellites were safely deployed, restoring faith in the Payload Assist Module and revitalising the Shuttle's own reputation. A concertina-like array of solar panels was deployed and retracted several times. With an overall length of 102 metres, it was the largest structure ever deployed from a U.S. spacecraft. Payload specialist Charles D. Walker gave a clear demonstration of the importance of having a trained specialist available on the spot when he repaired the malfunctioning continuous flow electrophoresis experiment, which otherwise would have failed. This system separates out a high-purity pharmaceutical product in the gravity-free environment of the spacecraft in very much greater quantities and at much higher levels of purity than can be done on Earth, where gravity causes mixing problems. Sponsors McDonnell Douglas and Johnson & Johnson intend to start human trials of the drug in 1985. The nature of the drug will not be divulged until the trials begin, but the sponsors expect to generate an annual turnover of a billion dollars in the 1990s from this commodity alone. Other potential space-produced pharmaceuticals may make this one of the most lucrative growth areas in space processing.

One curious problem that arose on the mission was the growth of substantial icicles from two waste water ports above the leading edge of the port wing. The main icicle – about 0.5 metres in length – was eventually dislodged by the manipulator arm to the relief of all concerned, since there had been fears that had the icicle broken off during reentry it could have damaged one of the orbital manoeuvring engine pods. Indeed, a similar problem is believed to have damaged Challenger during its reentry in the previous February.

The Shuttle has had its share of problems, its schedules have slipped, and the planned annual launch rate has yet to be attained; nevertheless, it has proved itself to be a practical and flexible system. When the fourth orbiter, 'Atlantis', enters service in 1985, NASA's fleet will be complete.

The Soviet Union is pressing ahead in developing its own Shuttle, similar in its major elements to the American vehicle but possibly having a higher payload capability. Tangible proof that the Soviet Union is also developing a smaller winged spaceplane came with the flight of what is believed to have been a 900-kilogram scale model of what will be an 18-tonne vehicle, on 3 June 1982. Launched from Kapustin Yar, the craft made a single orbit and splashed down south of the Cocos Islands 109 minutes after lift-off. Photographs taken after splashdown by the Royal Australian Air Force showed that the model spacecraft was 4–5 metres long and had the form of a lifting body, which blended into two steeply angled wings. Two further orbital tests took place in 1983. The full-size spaceplane would presumably be used to transport men and materials to and from an orbiting space station, or to inspect satellites; clearly, it could also have an important military role.

Large Space Stations

The Soviet Union is completing the development of a launch vehicle comparable to but probably more powerful than the Saturn 5. With a payload capacity to low Earth orbit of about 150 tonnes, this booster will probably be used to launch a large space station either as a single unit or in segments that would be assembled in orbit. A permanently manned Soviet space station, capable of housing 12–20 cosmonauts, is likely to be operational before the end of this decade.

In January 1984 President Reagan gave the go-ahead to an eight-year, 8-billion-dollar programme to set up a permanently manned structure capable of housing 6–8 astronauts, with crews rotating every three to six months. There were echoes of the 'Kennedy declaration' in his State of the Union Address when he announced, 'Tonight I am directing NASA to develop a permanently manned space station, and to do it within a decade'. As presently envisioned, the station would weigh some 36 tonnes and would be centred on five pressurised modules – each about 7 by 5 metres – providing living quarters, a laboratory, and support facilities. To accommodate the great array of solar panels needed to supply some 120 kilowatts of power, the station structure would measure 60–90 metres across its maximum diameter. The station would be accompanied by several free-flying platforms and would probably include a servicing bay for satellites and free-flying platforms, which would be collected and redeployed by a form of space tug. Europe, Canada, and Japan are all expected to become involved in the project.

Although objections have been raised to the present concept, it seems certain to go ahead and should be operational by about 1992.

Conventional Launch Vehicles

In the realm of conventional launchers, the redoubtable Delta kept up its reliability record, but the European Ariane suffered a serious setback when, on its first commercial mission on 10 September 1982, the third-stage motor failed half-way through its scheduled burn and its payload of two communications satellites was lost. The problem was eventually diagnosed as being due to a turbopump that failed because the lubricating oil was not up to the job. Nine months were to elapse before the next Ariane launch, which successfully deployed two satellites. Further successful missions in October 1983 and March and May 1984 were followed in August of that year by the first launch of the updated Ariane–3, a stretched version with two strap-on boosters. This was the first launch to be the complete responsibility of the commercial marketing and operating company, Arianespace, which can legitimately claim to be the world's first truly commercial launcher company. With five successful launches out of five since the abortive Ariane L–5 mission, the vehicle seems to be well established as a viable contender in the launcher market and has some thirty firm bookings for its services.

China, Japan, and India all achieved successful launches during this period to emphasise further that the space business is no longer the sole prerogative of the U.S.A. and U.S.S.R.

Prospects for private organizations getting into the launcher business were encouraged when Space Services Inc., a company headed by ex–NASA astronaut Donald K. Slayton, fired their first 'Conestoga' rocket to a height of 314 kilometres in September 1982. Another step towards the low-cost private enterprise launcher – admittedly on a smaller scale – was the test-firing from the Pacific Ocean of Starstruck Inc.'s Dolphin rocket.

Satellites

The launching of a wide variety of satellites continued apace. During 1982, the Soviet Union's tally of unmanned launches passed the 1,500 mark, and its total of manned and unmanned launches for the years 1982 and 1983 were 102 and 98 respectively – a launch rate far higher than that of America.

Among the more interesting American satellites was Landsat 4, a more sophisticated version of the earlier Earth resources satellites. In addtion to the well-tried multispectral scanner, Landsat 4 carries an even more powerful device known as the thematic mapper, which acquires data at seven colour bands (rather than four), thus allowing better discrimination of the nature of surfaces and ground cover. Finer details were revealed by its superior (30-metre) resolving power.

Sadly, the spacecraft suffered a series of failures that severely degraded its performance. Broken wires reduced the power supply from its solar panels by 50 per cent, the main command and data handling computer failed, and the direct data link from the thematic mapper to the ground also failed. Although a follow-up spacecraft, Landsat 5, was launched in 1984, Landsat 4 remains a prime target for a repair and refuelling mission. Following the success of the Solar Max rescue, NASA is considering a Landsat rescue as a possibility for 1986.

Another interesting development was the inauguration of an international search and rescue satellite system (COSPAS/SARSAT). First into action was a Soviet COSPAS satellite, which notched up its first success on 10 September 1982 when it pinpointed the location of an emergency position-indicating beacon (EPIRB) on a crashed aircraft in British Columbia, Canada, and so led to the rescue of three crew members. By mid-1984 the number of lives saved by the system had exceeded two hundred, and, in a bizarre incident, the detection of an EPIRB led to the arrest and conviction of a person who had stolen it and inadvertently left it switched on.

From the astronomers' point of view the most notable event was the successful launching of the Infrared Astronomical Satellite on 26 January, 1983. The first satellite to be devoted to this important branch of astronomy, IRAS was a cooperative venture between NASA, the Netherlands, and the United Kingdom. In order to detect faint sources against the background 'noise' produced in the telescope system (the telescope itself, being a heat source, emits infrared radiation), the 0.57-metre aperture telescope was cooled to about 2°K (2 degrees above absolute zero) by liquid helium, and its operating lifetime ended when the helium ran out.

On its first day of operation it detected as many astronomical sources of infrared radiation as had previously been discovered in the entire history of Earth-based infrared astronomy. In its ten-month operating lifetime it carried out an all-sky survey and netted about 250,000 sources, measuring radiation with a sensitivity 1,000 times better than can be achieved on Earth, and at wavelengths that cannot be detected at all from ground level.

Among its discoveries were seven new comets, one of which became clearly visible to the naked eye. A surprising discovery was the presence of a shell or disc of solid particles around the bright star Vega. Similar discs were found around some 50 other stars, and some of these discs may represent planetary systems in the making. Many newborn stars were found shrouded in dense blankets or dust and, indeed, IRAS may have succeeded in locating all the star-formation sites in the Galaxy. Similar sites were identified in other neighbouring galaxies. A new class of remote galaxy, which emits a hundred times as much infrared as visible light, was identified, but it remains to be seen whether these are conventional galaxies shrouded in dust or some wholly new species to add to the cosmic 'zoo'.

The launch of IRAS was followed soon afterwards by a small Japanese X-ray satellite (ASTRO-B) and by the Soviet 'Astron' – an orbiting observatory, which carried a 0.8 metre aperture French ultraviolet telescope, currently the largest telescope in orbit—and an X-ray detector. In May 1983 the European Exosat, which, after delays, had been switched from Ariane to a Delta launch, was placed into a highly elliptical orbit that allowed it to use occultations of X-ray sources by the Moon to pin down accurately the positions of a large number of X-ray sources.

Exploration of the Solar System

Two Soviet spacecraft, Veneras 15 and 16, went into orbit

round Venus in October 1983 to begin a radar mapping programme. The radar scans, probing through the dense cloud cover, revealed craters, mountains, and depressions with a best resolution of about 2 kilometres, a significant improvement on the Pioneer Venus data.

Pioneer 10, the first spacecraft to make a close encounter with Jupiter, continues on its odyssey towards interstellar space. Together with its twin, Pioneer 11, it is conducting studies of the heliosphere, the region of space within which the solar wind blows and the solar magnetic field dominates over the tenuous field that permeates interstellar space. On 25 April 1983 Pioneer 10 attained a greater distance from the Sun than the *present* distance of Pluto, and on 13 June passed beyond the orbit of Neptune. It is now further from the Sun than any known planet. The motion of the two craft is being carefully monitored for deviations that might signify the presence of a dim and distant companion star to the Sun or, perhaps, a tenth planet.

During 1983 the International Sun-Earth Explorer (ISEE–3) was nudged away from its location between the Earth and Sun to begin a complex series of gravitational slingshots around the Moon. These culminated in a high-speed skim past the Moon, just 120 kilometres above its surface, which sent the spacecraft en route to an encounter with comet Giacobini-Zinner in September 1985. ISEE–3 is expected to pass through that comet's tail nearly six months before an assorted flotilla of European, Soviet, and Japanese spacecraft encounter that most famous of all comets – comet Halley.

Halley's comet, which was last seen in 1910, makes its next close approach to the Sun when it reaches perihelion on 9 February 1986. Because it will then be on the far side of the Sun from the Earth, it is unlikely to become a spectacular or even a conspicuous naked-eye object but, nevertheless, it will be the focus of an unprecedented international programme of Earth-based and space-based observations.

Japan is despatching two small (140 kg) spacecraft, one of which, Planet A, should pass within 150,000 kilometres of the comet's nucleus on 8 March 1986. The Soviet Union is sending two Venera-class probes, Vega 1 and Vega 2, both of which will fly by way of Venus in June 1985, ejecting atmospheric probes as they pass by. Vega 1 should pass about 10,000 km from the nucleus on 6 March 1986, and Vega 2 is expected to pass even closer three days later. Their cameras should resolve details as small as 180 metres across. As the comet's icy nucleus cannot be seen directly from Earth, its position will not be precisely known, but it is hoped that positional data obtained by the Vegas will allow last-minute corrections to be made to the trajectory of the European Giotto spacecraft, which, launched by Ariane, is due to reach the comet on 13 or 14 March. Giotto may then pass within 500 km of the nucleus and achieve a resolution of 50 metres. It is hoped that at least some of the five craft will get close enough to analyse the structure and composition of the comet and reveal the true nature of its nucleus before being incapacitated by collisions with solid lumps of cometary material.

The tale of the Galileo mission to Jupiter received a new twist in 1982 when approval was granted after all for the development of the existing Centaur vehicle as the high-energy upper stage for the Shuttle. The availability of Centaur, which is more powerful than the two-stage IUS, will allow Galileo to fly directly to Jupiter instead of following a convoluted series of gravitational slingshots. The probable launch date is May 1986 with arrival at Jupiter in September 1988. The Centaur upper stage will also be used for the delayed launch of the European International Solar Polar spacecraft, which will share the same launch window as the Galileo mission.

Another success for NASA was the reinstatement of a radar mapping mission to Venus. Although less sophisticated than the cancelled VOIR spacecraft, the new Venus Radar Mapper should be able to resolve details as small as 1 kilometre on the Venusian surface at about half the cost of the original proposal. Currently the VRM is scheduled for a Shuttle-Centaur launch in 1988.

NASA now intends to mount a series of carefully selected missions using low-cost spacecraft. Following Galileo and VRM, the most favoured missions are: a Mars orbiter, to examine the geochemistry and meteorology of the planet; a comet rendezvous or asteroid fly-by; and a Titan fly-by, to look in more detail at this strange Saturnian moon and its dense atmosphere.

Military Activity

During the past few years there have been ominous indications of an imminent escalation in military activity in space, particularly with regard to the development of space-based weapons systems. Symbolic of the increasing military significance of space was the establishment, on 1 September 1982, of a U.S. Air Force 'Space Command'; part of its role would be to counter Soviet advances in space systems and weapons.

Air-to-space antisatellite missiles – launched from aircraft and capable of reaching geosynchronous satellites – are currently being developed; for example, a two-stage missile of this type was test-fired from a McDonnell Douglas F-15 aircraft early in 1984. During the next few years, the United States Department of Defense expects to receive a major increase in its research and development budget for antisatellite systems – in particular, high-powered lasers.

In March, 1983, in his so-called 'Star Wars' speech, President Reagan advocated a massive research effort to develop space-based laser and particle-beam weapons capable of destroying intercontinental ballistic missiles in flight and so, perhaps, rendering obsolete those particular weapons of mass destruction. On the face of it, the concept of orbital 'battle stations' capable of giving sufficient continuous cover to knock out thousands of missiles in a few minutes would seem to belong in the realm of science fiction. Even if the prodigious technical problems could be overcome and the vast energy requirements could be met, the development and constructional costs would be crippling. Yet so much has happened during the Space Age that would have been regarded as pure science fiction a few decades ago that it would be unwise to dismiss the possibility out of hand.

More conventional weapons systems, such as orbiting

antisatellite (ASAT) and anti-ballistic-missile missile platforms could be established much more easily. Indeed, according to reports in *Aviation Week and Space Technology,* test-firing of ASAT missiles may already have taken place from the Cosmos 1267 module, which was attached to the Salyut 6 space station.

The U.S. Department of Defense is supporting a project to develop an orbital gun to defend U.S. spacecraft against killer satellites. One possible device is the electromagnetic railgun, which converts electrical power into magnetic pressure and spits out 'bullets' at colossal speeds. In ground-based experiments tiny projectiles have been fired at speeds of up to 8.6 kilometres per second, and it is anticipated that by about 1990 an operational system could be developed with the ability to fire projectiles at 100 km per second and to destroy intercontinental ballistic missiles in flight. A system of this kind could become a practical reality considerably sooner than laser or beam weapons.

Western military analysts consider that the Soviet winged spaceplane could be used for 'quick response' manned military missions and could, for example, inspect or destroy Western satellites. Similar opinions have been expressed in the Soviet Union concerning the possible role of the U.S. Space Shuttle. Modern military communications and surveillance rely heavily on satellites, and the vulnerability of these satellites has prompted the United States Air Force to advocate the development of a spaceplane capable of taking off from and landing on a conventional runway and 'scrambling' at short notice to counter possible Soviet threats or to replace disabled satellites in time of crisis.

Much has happened in the brief time that has elapsed since the 25th anniversary of Sputnik 1. Shuttle missions have become everyday events, and the construction of permanently manned space stations is imminent. Communications and applications satellites continue to play an increasing role in everyday life, and the first fruits of space materials processing are about to become commercially available. Space probes continue to explore the Solar System, and astronomical satellites have discovered yet more intriguing objects and phenomena in the universe around us. As mankind's involvement with space continues to grow, it is perhaps inevitable that the military dimension should also be escalating. Without question this remains an exciting but critical era in human history.

Appendix

Highlights of Space Activity

1957-1984

This list of *selected* highlights includes notable 'space firsts' relating to satellites, lunar probes, and manned spaceflights, and also includes all successful planetary spacecraft (together with some important failures), and all the manned Apollo missions.

Dates quoted in column 1 are for the launching of the spacecraft or satellite in question unless otherwise indicated. With long-duration planetary missions where the planetary encounter occurs in a different year from the launching of the spacecraft, the spacecraft is listed under the date of launching *and* under the date of the planetary encounter.

Date	*EVENT*
1957	
4 Oct.	Launching of *Sputnik 1* by U.S.S.R. First artificial satellite. Weighed 84 kg. Re-entered atmosphere and burned up on 4 Jan. 1958.
3 Nov.	*Sputnik 2*. Second artificial satellite. Weighed 500 kg. First to carry a living creature – the dog 'Laika'.
1958	
31 Jan.	*Explorer 1*. First American satellite. Weighed 14 kg including empty fourth stage. Reached maximum altitude of 2548 km; discovered inner Van Allen radiation belt.
18 Dec.	*Atlas–Score* (U.S.A.): broadcast first voice message from space.
1959	
2 Jan.	*Luna–1* (U.S.S.R). First lunar fly-by (4 Jan.)
12 Sep.	*Luna–2* (U.S.S.R.). First spacecraft to hit the Moon (on 13 Sept.)
4 Oct.	*Luna–3* (U.S.S.R.). First photographs of far side of Moon taken on 7 October and relayed to Earth on 18 Oct.

1960

1 Apr. *Tiros–1* (U.S.A.). First purpose-built meteorological satellite.

13 Apr. *Transit 1B* (U.S.A.). First navigation satellite.

12 Aug. *Echo 1* (U.S.A.). First passive communications satellite – 30-metre diameter, inflated orbiting balloon.

1961

12 Feb. *Venera 1* (U.S.S.R.). Fly-by of Venus at range 100,000 km on 19 May, but communication lost on 27 Feb.

12 Apr. *Vostok 1* (U.S.S.R.). Yuri Gagarin becomes first man in space. 108-minute orbital flight.

5 May *Mercury 3* (U.S.A.). Alan Shepard becomes first American to go into space on sub-orbital 'hop' in space capsule 'Freedom 7'.

1962

20 Feb. *Mercury 6* (U.S.A.). John Glenn becomes first American to go into orbit. 3 orbits completed in capsule 'Friendship 7'.

23 Apr. *Ranger 4* (U.S.A.). Crashes on to far side of Moon on 26 Apr.

10 Jly. *Telstar* (U.S.A.). First privately constructed active communications satellite. Relayed first live transatlantic TV programmes on 23 July.

27 Aug. *Mariner 2* (U.S.A.). First successful Venus fly-by on 14 December. Range 34850 km. Revealed temperature of over 400°C.

1 Nov. *Mars–1* (U.S.S.R.). Fly-by of Mars at range 193,000 km on 19 June, 1963 but contact lost 21 February.

1963

16 Jun. *Vostok 6* (U.S.S.R.). First woman to make a space flight—Valentina Tereshkova.

26 Jly. *Syncom II* (U.S.A.). First geosynchronous communications satellite.

1964

28 Jly. *Ranger 7* (U.S.A.). First successful Ranger transmits 4316 TV pictures prior to impact on Moon on 31 July.

12 Oct. *Voskhod 1* (U.S.S.R.). First three-man spacecraft.

28 Nov. *Mariner 4* (U.S.A.). First successful Mars fly-by achieved on 14 July, 1965 at range 9850 km. 21 photographs revealed Martian craters.

1965

18 Mar. *Voskhod 2* (U.S.S.R.). First EVA ('spacewalk') carried out by Alexeï Leonov.

23 Mar. *Gemini 3* (U.S.A.). First manned Gemini craft. First to carry out orbital manœuvres.

23 Apr. *Molniya 1* (U.S.S.R.). First operational Soviet communications satellite in 12-hour orbit.

16 Nov. *Venera 3* (U.S.S.R.). First spacecraft to strike Venus (on 1 March, 1966) but contact lost – no data received.

26 Nov. France becomes first nation other than U.S.A. and U.S.S.R. to place a satellite in orbit by means of its own launcher: A–1 satellite weighing 40 kg by means of *Diamant* launch vehicle.

15 Dec. *Gemini 6 and 7* (U.S.A.). Achieve first rendezvous in space: approach to within 2 metres.

1966

31 Jan. *Luna 9* (U.S.S.R.). First soft-landing on Moon achieved on 3 February. First photographs from lunar surface.

16 Mar. *Gemini 8* (U.S.A.). First docking with orbiting target (Agena). Malfunctioning thruster caused tumbling of spacecraft. Emergency landing achieved.
31 Mar. *Luna 10* (U.S.S.R.). First lunar satellite.
30 May *Surveyor 1* (U.S.A.). First American soft-lander on Moon; touchdown on 2 June.
10 Aug. *Orbiter 1* (U.S.A.). First of series of American photographic orbiters for lunar mapping.

1967
27 Jan. Fire in launchpad test of Apollo spacecraft results in deaths of astronauts Grissom, White, and Chaffee.
23 Apr. *Soyuz 1* (U.S.S.R.). First manned flight of the Soyuz series spacecraft. Parachute lines became tangled on re-entry. Cosmonaut Komarov killed in crash landing.
12 Jun. *Venera 4* (U.S.S.R.). Entered atmosphere of Venus on 18 October and continued to transmit data until an altitude of 26 km above the Venusian surface.
14 Jun. *Mariner 5* (U.S.A.). Successful Venus fly-by on 19 October.

1968
14 Sep. *Zond 5* (U.S.S.R.). First unmanned spacecraft to fly round the Moon and return to the Earth.
11 Oct. *Apollo 7* (U.S.A.). First manned Apollo test flight in Earth orbit.
20 Oct. *Kosmos 249* (U.S.S.R.). Believed to be first orbital test of a 'killer' satellite.
7 Dec. *OAO 2* (U.S.A.). First successful major unmanned Orbiting Astronomical Observatory; ultraviolet studies of stars.
21 Dec. *Apollo 8* (U.S.A.). First manned lunar orbital mission. Frank Borman, James Lovell, and William Anders become first men to go into orbit round Moon on 24 December.

1969
5 Jan. *Venera 5* (U.S.S.R.). Entered Venusian atmosphere on 16 May.
10 Jan. *Venera 6* (U.S.S.R.). Entered Venusian atmosphere on 17 May.
3 Mar. *Apollo 9* (U.S.A.). First test of Lunar Module in Earth orbit.
15 Feb. *Mariner 6* (U.S.A.). Successful Mars fly-by on 31 July.
26 Mar. *Meteor 1* (U.S.S.R.). First operational meteorological satellite of the Soviet 'Meteor' network.
27 Mar. *Mariner 7* (U.S.A.). Successful Mars fly-by on 5 August.
18 May *Apollo 10* (U.S.A.). First test of Lunar Module in lunar orbit.
16 Jly. *Apollo 11* (U.S.A.). First manned landing on the Moon achieved on 20 July in Sea of Tranquillity. Neil Armstrong followed by Edwin Aldrin become first men to set foot on the Moon. Michael Collins remained in orbit in Command Module.
14 Nov. *Apollo 12* (U.S.A.). Second manned lunar landing.

1970
11 Feb. Japan becomes second nation apart from U.S.S.R. and U.S.A. to launch a satellite ('*Osumi*' – 24 kg) by means of its own launch vehicle – *Lambda-4S*.
11 Apr. *Apollo 13* (U.S.A.). Service Module oxygen tank explosion on outward flight towards Moon cripples spacecraft. By using Lunar Module systems, crew safely returned to Earth in first deep-space emergency.
24 Apr People's Republic of China becomes third nation apart from U.S.S.R. and

U.S.A. to launch a satellite (*China 1*; 173 kg) with own ICBM-based launch vehicle.

17 Aug. *Venera 7* (U.S.S.R.). First successful soft-landing of capsule on Venus, on 15 December. Transmitted from surface for 23 minutes.

12 Sep. *Luna 16* (U.S.S.R.). Landed on Moon on 20 September, drilled core sample and returned this to Earth on 24 September. First remote-sample return mission.

10 Nov. *Luna 17* (U.S.S.R.). Touchdown on Moon on 17 Nov with *Lunokhod 1*. First remote-controlled lunar roving vehicle.

12 Dec. *Explorer 42* ('*Uhuru*') (U.S.A.). First purpose-built X-ray astronomy satellite.

1971

31 Jan. *Apollo 14* (U.S.A.). Third manned lunar landing.

19 Apr. *Salyut 1* (U.S.S.R.). First orbital space station. Re-entered atmosphere and broke up on 11 October.

19 May *Mars 2* (U.S.S.R.). Entered orbit round Mars on 27 November. First capsule to reach martian surface – crashed; no data.

28 May *Mars 3* (U.S.S.R.). Entered orbit round Mars on 2 December. Capsule landed but transmitted for only 20 seconds.

30 May *Mariner 9* (U.S.A.). First successful Mars Orbiter – entered orbit 14 November. Returned 6876 pictures.

6 June *Soyuz 11* (U.S.S.R.). 23-day visit to Salyut 1 space station. Decompression of spacecraft during re-entry caused deaths of cosmonauts Dobrovolsky, Volkov, and Patsayev.

26 Jly. *Apollo 15* (U.S.A.). Fourth lunar landing. First Apollo mission to use the Lunar Rover.

28 Oct. Great Britain becomes fourth nation other than U.S.S.R. and U.S.A. to launch its own satellite ('*Prospero*'; 66 kg) by means of its own launch vehicle (*Black Arrow*).

1972

2 Mar. *Pioneer 10* (U.S.A.). First fly-by of Jupiter achieved on 3 December 1973 at a range of 130,345 km. Wealth of data returned. First spacecraft to exceed escape velocity of Solar System.

27 Mar. *Venera 8* (U.S.S.R.). Landed on Venus on 22 July; transmitted for 107 min.

16 Apr. *Apollo 16* (U.S.A.). Fifth manned lunar landing.

27 Jly. *Landsat 1* (ERTS 1) (U.S.A.). First Earth resources surveying and monitoring satellite.

7 Dec. *Apollo 17* (U.S.A.). Sixth and final manned landing of the Apollo series.

1973

5 Apr. *Pioneer 11* (U.S.A.). First Jupiter-Saturn mission. Jupiter fly-by on 3 December, 1974 and first Saturn fly-by on 1 September, 1979 at range 21,000 km.

14 May *Skylab* (U.S.A.). First, and so far only, American space station. Damage to shielding and solar panels sustained during launch.

25 May First manned expedition to Skylab. Crew of Conrad, Kerwin, and Weitz successfully repair space station.

21 Jly. *Mars 4* (U.S.S.R.). Mars fly-by on 10 February, 1974.

25 Jly. *Mars 5* (U.S.S.R.). Entered orbit round Mars on 12 February, 1974.

5 Aug. *Mars 6* (U.S.S.R.). Capsule landed on Mars but contact lost prior to touchdown.

4 Nov. *Mariner 10* (U.S.A.). First spacecraft to complete a two-planet mission;

fly-by of Venus on 2 February, 1974, and first fly-by of Mercury on 29 March, 1974.

3 Dec. *Pioneer 10* (U.S.A.). First Jupiter fly-by (launched 2 Mar. 1972).

1974

5 Feb. *Mariner 10* (U.S.A.). Makes fly-by of Venus. First cloud cover photographs (launched 4 Nov., 1973).

29 Mar. *Mariner 10* (U.S.A.). Makes first of three Mercury encounters.

30 May *ATS–6* (U.S.A.). First experimental communications satellite with capability for direct broadcasts to small domestic antennæ.

3 Dec. *Pioneer 11* (U.S.A.). Jupiter fly-by (launched 5 Apr., 1973).

10 Dec. *Helios 1* (NASA/West Germany). First spacecraft to approach as close as 45 million km to the Sun.

1975

6 Jun. *Venera 9* (U.S.S.R.). First picture of surface of Venus obtained after landing on 22 October.

14 Jun. *Venera 10* (U.S.S.R.). Landed on Venus on 25 October. Picture transmitted.

17 Jly. *Apollo-Soyuz Test Project*. Joint American-Soviet link up in orbit symbolized when astronaut Stafford shakes hands with cosmonaut Leonov through docking tunnel.

20 Aug. *Viking 1* (U.S.A.). First successful Mars lander. Entered orbit round Mars on 19 June, 1976; lander craft touched down on 20 July, 1976. Lander still returning data in 1982.

9 Sep. *Viking 2* (U.S.A.). Second successful Mars lander. Entered orbit round Mars on 7 August, 1976; touchdown of lander on 3 September, 1976.

1976

20 Jly. *Viking 1* (U.S.A.). First successful Mars landing (launched 20 Aug., 1975).

9 Aug. *Luna 24* (U.S.S.R.). Landed on Moon and returned soil sample to Earth. Last lunar spacecraft to date.

3 Sep. *Viking 2* (U.S.A.). Second successful Mars landing (launched 9 Sep., 1975).

1977

12 Aug. First approach and landing test free flight of the *Space Shuttle* 'Enterprise' (U.S.A.).

20 Aug. *Voyager 2* (U.S.A.). Jupiter fly-by 9 July, 1979, Saturn fly-by 26 August, 1981; currently *en route* for Uranus (1986) and Neptune (1989) encounters.

5 Sep. *Voyager 1* (U.S.A.). Jupiter fly-by 5 March, 1979, Saturn fly-by 12 November, 1980. First close-up views of satellite Titan.

29 Sep. *Salyut 6* (U.S.S.R.). Most successful Salyut space station to date. Visited to end 1981 by 16 manned spacecraft and 12 unmanned supply ships.

1978

20 Jan. *Progress 1* (U.S.S.R.). First unmanned 'cargo' spacecraft to dock with Salyut space station to replenish supplies.

20 May *Pioneer Venus 1* (U.S.A.). Entered orbit round Venus 4 December. Radar maps of planetary surface.

8 Aug. *Pioneer Venus 2* (U.S.A.). 5 probes entered venusian atmosphere on 9 December; probes reached surface.

9 Sep. *Venera 11* (U.S.S.R.). Landed on Venus on 25 December and transmitted for 95 minutes.

14 Sep. *Venera 12* (U.S.S.R.). Landed on Venus on 21 December; transmitted for 110 minutes.

1979

5 Mar.	*Voyager 1* (U.S.A.). Jupiter fly-by (launched 5 Sep., 1977): Discovery of Jupiter's ring, Io's volcanoes.
9 Jly.	*Voyager 2* (U.S.A.). Jupiter fly-by (launched 20 Aug., 1977).
11 Jly.	Re-entry and destruction of *Skylab* (U.S.A.).
1 Sep.	*Pioneer 11* (U.S.A.). First Saturn fly-by (launched 5 April, 1973).
24 Dec.	*Ariane* (European Space Agency). First successful launching of a test satellite by the European launch vehicle, Ariane.

1980

9 Apr.	*Soyuz 35* (U.S.S.R.). Cosmonauts Popov and Ryumin enter Salyut 6 to commence a mission which lasted for 185 days.
18 Jly.	India becomes fifth nation apart from U.S.S.R. and U.S.A. to launch a satellite ('*Rohini*'; 35 kg) by means of own launch vehicle – *SLV–3*.
12 Nov.	*Voyager 1* (U.S.A.). Saturn-Titan fly-by (launched 5 Sep. 1977). Complexity of rings revealed. Titan shown to have nitrogen atmosphere.

1981

12 Apr.	*STS–1* (U.S.A.). First manned orbital flight of the American Space Shuttle 'Columbia', 2-day mission crewed by John Young and Robert Crippen. Launch from Cape Canaveral, landing on runway at Edwards Air Force Base, California. First flight of re-usable spacecraft.
26 Aug.	*Voyager 2* (U.S.A.). Saturn fly-by (launched 20 Aug. 1977).
30 Oct'	*Venera 13* (U.S.S.R.). Landing on Venus on 2 March, 1982; colour photographs of surface.
4 Nov.	*Venera 14* (U.S.S.R.) Landing on Venus on 5 March, 1982; colour photographs of surface.
12 Nov.	*STS–2* (U.S.A.). Second manned orbital mission of the Space Shuttle 'Columbia' and the first occasion on which a spacecraft has been used twice. 2-day mission crewed by astronauts Engel and Truly.

1982

22 Mar.	*STS–3* (U.S.A.). Third manned orbital flight of the Space Shuttle 'Columbia'. 8-day mission crewed by astronauts Lousma and Fullerton.
19 Apr.	*Salyut 7* (U.S.S.R.). Launching of new space station.
14 May	*Soyuz T-5* (U.S.S.R.). Cosmonauts Berezovoy and Lekedev transfer to *Salyut 7* to commence 211-day space mission.
27 Jun.	*STS-4* (U.S.A.). Fourth manned orbital flight of the Space Shuttle 'Columbia'. 7-day mission crewed by astronauts Mattingley and Hartsfield.
19 Aug.	*Soyuz T-7* (U.S.S.R.). Svetlana Savitskaya becomes second woman in space.
11 Nov.	*STS-5* (U.S.A.). First 4-man mission; fifth orbital flight of 'Columbia'. First Shuttle launch of commercial satellites.

1983

26 Jan.	IRAS (NASA/Netherlands/United Kingdom). First infrared astronomical satellite.
4 Apr.	*STS-6* (U.S.A.). First flight of Space Shuttle 'Challenger'. Four-man crew commanded by Paul J. Weitz. Failure of IUS upper stage during deployment of TDRS satellite.
20 Apr.	*Soyuz T-8* (U.S.S.R.). Fails to dock with *Salyut 7*. Returns to Earth on 22 April.
13 Jun.	*Pioneer 10* (U.S.A.) passes beyond the orbit of Neptune.

18 Jun. *STS-7* (U.S.A.). Second 'Challenger' mission. First 5-person crew, commanded by Robert L. Crippen. First American woman in space (Sally K. Ride).

27 Jun. *Soyuz T-9* (U.S.S.R.). Vladimir Lyakhov and Alexander Alexandrov enter *Salyut 7* on 28 June to commence 149-day stay.

30 Aug. *STS-8* (U.S.A.). Third orbital mission of 'Challenger'. Five-person crew commanded by Richard H. Truly. First night launch and landing of Shuttle.

9 Sep. Major fuel leak in *Salyut 7*. Repaired during EVAs carried out over next 11 months.

28 Sep. Soyuz booster (U.S.S.R.) explodes on launch pad. Cosmonauts Titov and Strekalov pulled clear by escape rockets.

28 Nov. *STS-9/Spacelab* (U.S.A./ESA). 'Columbia' with 6-man crew commanded by John Young takes *Spacelab* into orbit for the first time. First West German astronaut—Dr. Ulf Merbold.

1984

3 Feb. *Shuttle 41-B* (U.S.A.). 'Challenger', commanded by Vance Brand, makes tenth Shuttle mission. Two satellites fail to reach required orbits due to failures of Payload Assist Modules. First free-flight of Mannen Manoeuvring Units by astronauts McCandless and Stewart.

8 Feb. *Soyuz T-10* (U.S.S.R.). Cosmonauts Vladimir Solovyov, Leonid Kizim and Oleg Atkov enter Salyut 7 on 9 February to commence record-breaking space endurance mission.

3 Apr. *Soyuz T-11* (U.S.S.R.). Three-man visit to *Salyut 7*. First Indian cosmonaut: Rakesh Sharma.

6 Apr. *Shuttle 41-C* (U.S.A.). 'Challenger' commanded by Robert L. Crippen. Rescue and repair of Solar Max satellite.

17 Jly. *Soyuz T-12* (U.S.S.R.). Three-person visit to *Salyut 7*.

30 Aug. *Shuttle 41-D* (U.S.A.). First flight of orbiter 'Discovery', commanded by Henry W. Hartsfield. Three communications satellites successfully deployed.

A Guide to Further Reading

A great many books have been written on various aspects of space exploration, but the following may be of particular interest to readers of this book wishing to pursue the subject further.

The Illustrated Encyclopaedia of Space Technology, edited by Kenneth Gatland. Salamander Books, London, 1981.
An authoritative, detailed, and splendidly illustrated reference book on all aspects of the subject.

The Rocket, by David Baker, New Cavendish Books, London, 1978. A comprehensive history of rockets and missiles.

The History of Manned Spaceflight, by the same author and publisher, 1981.
A comprehensive survey of manned spaceflight.

Contrasting views of the Soviet space programme are to be found in two readable and interesting accounts:

Soviets in Space, by Peter Smolders, Lutterworth Press, London, 1973.
Red Star in Orbit, by James Oberg, Harrap, London, 1981.

Comprehensive coverage of Soviet unmanned and manned programmes is to be found in two volumes by Nicholas L. Johnson:

Handbook of Soviet Lunar and Planetary Exploration, and
Handbook of Soviet Manned Spaceflight, published in 1979 and 1980 by Univelt, Inc.

One of a number of books exploring the theme of the political background to the Apollo project is:

The Decision to Go to the Moon – Project Apollo and the National Interest, by John M. Logsdon, The M.I.T. Press, Cambridge, Mass., and London, 1970.

The Versatile Satellite, by Richard W. Porter, Oxford University Press, 1977. Gives a concise but thorough account of the various applications to which satellites can be put.

Planetary Encounters, by Robert M. Powers, Sidgwick & Jackson, London, 1982. Traces history and development from pre-Christian era to present day.

The Promise of Space, by Arthur C. Clarke, Hodder and Stoughton. London, 1968.
Although written prior to the Apollo lunar landing, this book by one of the outstanding popularizers of Space exploration and 'father of the geostationary communications satellite' remains a fascinating account of the history of and future prospects for the exploration and exploitation of space.

Update on Space, Vol. 1, edited by B. J. Bluth and S. R. McNeal, National Behaviour Systems, Granada Hills, California, 1981.
An interesting collection of essays on the future exploitation of space and the likely human benefits.

The Road to the Stars, by Iain Nicolson, Westbridge Books (a Division of David & Charles), Newton Abbot, 1978.
A look at the prospects for interstellar flights.

The View from Space, by Merton E. Davies and Bruce C. Murray, Columbia University Press, 1971. Surveys the scope and techniques for studying the Earth and planets from space.

Earthwatch, by Charles Sheffield, Sidgwick & Jackson, London, 1981.
Vividly illustrated example of satellite images with informative text.

NASA publications, obtainable from the Superintendent of Documents, U.S. Government Printing Office, Washington, D.C. 20402, cover a wide range of topics, including 'Chronologies' of specific American space programmes, such as Mercury, Gemini, Apollo, Skylab, etc., and more general historical aspects of rocketry and spaceflight.
NASA Press releases (NASA News) and Press Kits are an invaluable source of up-to-date information.

Up-to-date information is obtainable in two particularly excellent magazines:

Aviation Week and Space Technology, McGraw-Hill, New York (weekly) and *Spaceflight,* British Interplanetary Society, 27/29 South Lambeth Rd., London SW8 1SZ (monthly). The Society's publication *Space Education* is also of great value to the general reader.

Index

The symbol '(l.v.)' denotes that a name refers to a launch vehicle. A page number in italic signifies an illustration.

A-series (l.v.), 46, 48, 122, 126, 131
aerospace plane, 170, 188
Aldrin, Edwin, 65, 133, 138, 144, 146
aphelion, 21
apogee, 30-1
Apollo Project, 131, 135, 137-153
Apollo 8, 33, 142-4
Apollo 11, 65, 137, 144, *145*
Apollo 13, 148-9
Apollo-Soyuz Test Project (ASTP), 163-4
Ariane (l.v.), 50, 178-9, 203
Ariel, 103
Armstrong, Neil, 65, 132, 137-8, 144-6, 153
asteroids, 22, 86-7, 184-5
Atlas (l.v.), 48, 60, 126
Atlas-Score, 48, 108

Baikonur Cosmodrome, 12, 122
Big Bird, 120
black hole, 25-6, 103-4
Borman, Frank, 132, 143
British Interplanetary Society, 41, 141, 191

Cape Canaveral, 14, 33, 127
Centaur (l.v.), 48-9, 60, 83, 177, 183
circular velocity, 30, *31*, 70
Clarke, A. C., 107, 112
combustion chamber, 35, 40
Collins, Michael, 144
communications satellite, 106-112, 181-2
comet, 22-3, 30, 184
Congreve, William, 38-9
Conrad, Charles, 148, 160
Cosmos series, 97-8, 120-1, 166, 194
Crippen, Robert, 173, 199-200
Cygnus X-1, 103-4

Deep Space Network, 75, 87
Delta (l.v.), 49, 203
Diamant (l.v.), 49, 197
Discoverer series, 119
D-series (l.v.), 47
Dyna-Soar, 170

Earth, the, 18-21, 24
Echo I and II, 108
electromagnetic radiation, 94-6
Elektron series, 98
ellipse, 30, *31*
escape velocity, 29, *31,* 32, 70
European Space Agency (ESA), 50
exhaust velocity, 35-6
Explorer series, 97, 99, 101-2
Explorer 1, 15, 97

Gagarin, Yuri, 122, *123,* 124
Galileo, 182, 207
Gemini series, 48, 129, 131-3, *134,* 135
Giotto, 184
Glenn, John, 127-8
gravitation, 28-9
gravitational 'slingshot', 73-5, 183
G-series (l.v.), 47, 152, 167

HEAO, 103
Hohmann, Walter, 71
Helios, 100

ICBM, 45
industrial activity in space, 166, 187
Inertial Upper Stage, 177-8, 182-3
Intelsat, 109-10
ISPM, 183
interstellar flight, 190-1
Io, 89-90
ionosphere, 107

Juno-1 (l.v.), 15, 139
Jupiter, 21, 24, 86-91
  Great Red Spot, 86, 89, *90*
Jupiter-C (l.v.), 14, 47, 139

Kapustin Yar, 43
Kennedy, President John F., 125, 137
Kerwin, Joseph, 160
'killer' satellite, 121
Komarov, Vladimir, 129, 135-6, 155
Korolev, Sergei, 11, 41, *123*

Laika, 13
Landsat (ERTS), 116-7, 204
launch window, *73*
Leonov, Alexei, 130, 164
light-year, 23
Lovell, James, 132-3, 143, 148
Luna series, 53-60, 64-5
lunar month, 51, *52*
lunar orbital rendezvous, 138
lunar roving vehicle, 149-50
Lunokhod, 65, *66,* 67
L-sat, 111, *112*

manned missions, future, 189-90
Mariner series, 69, 75, 77-8, 81-3, 85-6
Mars, 21, 24, 80-5
Mars series, 81-3
mass ratio, 35
Mercury, 21-2, 24, 85-6
Mercury Project, 126-9
meteorological satellites, 113-6
Meteor series, 116
military satellites, 119-21

Moon, the, 19, 51-3, 150
race, 152-3
spaceprobes to, 51-68
Molniya series, 110-11

NASA, 126
navigational satellites, 118-9
Neptune, 21, 24
Newton, Isaac, 28, 30
NOAA satellites, 115-6

Oberg, James, 153
Oberth, Hermann, 41, 154
O'Neill, Prof. Gerard K., 89
orbital transfer system, 188
Orbiter, Project, 13, 14
Orbiter series, 62, *63*, 64
Orbiting Astronomical Observatory (OAO), 102-3

parking orbit, 32
perigee, 12, 30, *31*
perihelion, 21
Pioneer series, 54-5, 57, 75, 79, 86-8, 206
planetary missions, future, 184-5
Pluto, 21, 24
Popov, Leonid, 167
Progress spacecraft, 165, 167, 193, 195
propellant, 35, 131
Proton (l.v.), 156 (see: D-series)

Ranger series, 57-9
Redstone (l.v.), 13-14, 47, 126
rocket, principle of, 34
types of, 35-6, *37*, 38, 40, 43, 45, 90-1
Ryumin, Valeri, 167

Salyut series, 156, *157*, 158, 164-7, 193-4
satellite solar power stations (SPS), 187-8
Saturn, 21, 24, 66, 89, 91-3, *92*
Saturn (l.v.), 36, 47, 49, 139-40, 143-5
Scout (l.v.), 49
Shepard, Alan, 127
Skylab, 100, 159-161, *162*, 163, 214
Solar System, 20-3
solar wind, 97, 99, 101
Solar Maximum Mission, 100-1, 199-200
Soviet Shuttle, 179
Soyuz 1, 135, 155
Soyuz series, 155-8, 164-6, 194-5
Soyuz-T, 155-6, 166
space colonies, 188-9
space endurance record, 167
Spacelab, 176-7, 197
Space Shuttle, 49, 101, 168-174, *175*, 176-80, 191, 196-201
space station, 154, 167, 169, 187, 202
Space Telescope, 177, 185-6
Sputnik 1, 9, 11, 16-7, 97, 106, 211
Sputnik 2, 10-14
Sputnik 3, 15-6
SS-6 (ICBM), 45
stars, 23-6
step-rocket, 35-7
Sun, the, 18-20, 24

Sun-synchronous orbit, 113-4, *115*
Surveyor series, 60-2, 148
Syncom, 109

Telstar, 108-9
Tereshkova, Valentina, 129
Tiros, 113-4
Titan (l.v.), 48-9, 83, 131
Titov, Hermann, 125
thrust, 35, 42
Transit 1B, 118
transfer orbit, 71, 72
Tsiolkovsky, Konstantin E., 39-40, 154

Uranus, 21, 24
Uhuru, 98, 103

V-2, 42-3, *44,* 45
Van Allen zones, 15-6, 55, 87, 98
Vanguard, 14, 16, 48, 97
Venera series, 69, 77-80, 184, 207
Venus, 21, 24, 76-80
Verne, Jules, 33
Viking, 83-4, *85*
von Braun, Wernher, 13, 42-3, 47, 137, 152, 154, 168-9
Voskhod, 129-30
Vostok 1, 122, *124*
Vostok series, 122, *124,* 125, 129
Voyager, 76, 88-93, 182

Wac Corporal, 43
weightlessness, 29
Weitz, Paul, 160
White, Edward, 132, *133,* 135

X-ray astronomy, 96, 102-4
X-15, 169-70

Young, John, 132, 144, 173

Zond series, 64, 81, 143, 152,